Corrosion Control in the Chemical Process Industries

Second Edition

MTI Publication No. 45

C. P. Dillon

 **Materials Technology Institute
of the Chemical Process
Industries, Inc.**

Published for MTI by NACE International

About MTI

The Materials Technology Institute of the Chemical Process Industries, Inc. (MTI) is a unique, cooperative research and development organization representing private industry. Its objective is to conduct generic, nonproprietary studies of a practical nature on the selection, design, fabrication, testing, inspection, and performance of materials and equipment used in the process industries.

Through membership in MTI, companies can solve nonproprietary problems of major concern to the process industries, leverage research and development dollars by participating in the direction and results of MTI studies, capitalize on the expertise of member company representatives, and learn about MTI's accomplishments in a timely fashion.

More information about membership in MTI can be obtained from the Executive Director. MTI members receive copies of MTI publications as part of their membership.

**Materials Technology Institute of the
Chemical Process Industries, Inc.**

1215 Fern Ridge Parkway, Suite 116
St. Louis, MO 63141-4401
Tel: 314/576-7712
Fax: 314/576-6078

Contents

Tables xv

Figures xvii

Preface to Second Edition xix

Preface to First Edition xxi

Chapter 1

Introduction 1
   1.1 Purpose 1
   1.2 Cost of Corrosion 1
   1.3 Methodology 2
   1.4 Resources 4

Section I — Fundamental Factors

Chapter 2

Basic Considerations 15
   2.1 Safety and Reliability 15
   2.2 Cost 17
   2.3 Environmental Aspects 19
   2.4 Energy Considerations 20
   2.5 Materials Conservation 20

Chapter 3

Factors in Materials Selection 23
   3.1 Cost 23
   3.2 Physical Properties 23

3.3   Mechanical Properties 25
3.4   Codes and Regulations 29
3.5   Fabrication Characteristics 32
3.6   Corrosion Characteristics 42
3.7   Amenability to Corrosion Control 43

Chapter 4

Materials Selection Procedures 45
4.1   Materials Considerations 45
4.2   Materials-Environment Interactions 46
4.3   Specific Equipment 46
4.4   Procedures and Communications 47
4.5   Conclusion 50

Section II — Corrosion Factors

Chapter 5

Corrosion Mechanisms 55
5.1   Definition 55
5.2   Electrochemistry of Metallic Corrosion 55

Chapter 6

Corrosion and Metallurgical Phenomena 69
6.1   Corrosion Phenomena 69
6.2   Metallurgical Phenomena 74

Chapter 7

Sensitization and Weld Decay 77

Chapter 8

Environmental Cracking 83

8.1   Metallic Materials 83
8.2   Plastics 90

Chapter 9

Corrosion Testing 93
9.1   Materials Factors 93
9.2   Materials Characteristics 95
9.3   Laboratory Tests 96
9.4   Field Tests 98
9.5   Corrosion Coupons 101
9.6   Coupon Evaluation 101

Section III — Materials

Chapter 10

Light Metals 105
10.1 Structure of Metals 105
10.2 Magnesium 105
10.3 Aluminum and Its Alloys 107

Chapter 11

Iron and Steel 111
11.1 Cast Irons 111
11.2 Steels 114
11.3 Numbering 121

Chapter 12

Stainless Steels 123
12.1 Nature of Stainless Steel 123
12.2 Types of Stainless Steel 124
12.3 Stainless Castings 131

Chapter 13

High-Performance Nickel-Rich Alloys 133
   13.1 Superaustenitics 134
   13.2 High-Nickel Austenitics 136
   13.3 Ranking Alloys 136

Chapter 14

Nickel and Its Alloys 139
   14.1 Nickel Alloys 139
   14.2 Chromium-Bearing Alloys 142

Chapter 15

Lead, Tin and Zinc 145
   15.1 Lead and Its Alloys 145
   15.2 Tin 147
   15.3 Zinc 147

Chapter 16

Copper and Its Alloys 149
   16.1 Copper 149
   16.2 Brass 151
   16.3 Bronze 152

Chapter 17

Reactive, Refractory and Noble (Precious) Metals 155
   17.1 Reactive and Refractory Metals 155
   17.2 Noble (Precious) Metals 160

Chapter 18

Nonmetallic Materials 163

Chapter 35

Corrosion Inhibitors 339
   35.1 Definition 339
   35.2 Electrochemistry 339
   35.3 Aqueous Systems 340
   35.4 Refrigeration Brines 341
   35.5 Acids 342
   35.6 Acid-Gas Scrubbing Systems 342
   35.7 Nonaqueous Environments 343
   35.8 Other Considerations 343

Chapter 36

Paints and Coatings 345
   36.1 Temporary Rust Preventatives 346
   36.2 Paints and Coatings 346

Chapter 37

Anti-Corrosion Barriers 365
   37.1 Metallic Barriers 365
   37.2 Nonmetallic Barriers 369

Chapter 38

Cathodic Protection 371
   38.1 Definition 371
   38.2 Principles 371
   38.3 Design and Application 377
   38.4 Specialty Applications 378

Chapter 39

Inspection and Failure Analysis 381
   39.1 Inspection 381

39.2 Failure Analysis 384
39.3 Preventive and Predictive Maintenance 386
39.4 Documents and Data/Information Retrieval 387

Glossary of Acronyms 393

End Notes 397

Index 401

Tables

2.1   Specific Protective Measures 18

5.1   Metals 57
5.2   Electromotive Series 62
5.3   Galvanic Series in Seawater 63

8.1   Environmental Cracking Susceptibility 87

10.1   UNS Numbers: Aluminum Alloys..................... 108

12.1   Wrought Stainless Steels 126
12.2   Duplex Stainless Steels..................................... 130
12.3   Designations: Stainless Steel Castings 131

13.1   Fe-Ni-Cr-Mo-Cu Alloys 135
13.2   Fe-Ni-Cr-Mo Alloys (6% Mo Superaustenitics) . 135
13.3   Ni-Cr-Fe-Mo-Cu Alloys 137
13.4   Ni-Cr-Fe-Mo-Cu Alloys 137

14.1   Ni-Cr-Mo Alloys 143

16.1   Copper Alloys 150

17.1   Titanium Alloys 157

19.1   Characteristics: Zinc Water Analysis 184
19.2   Scaling by Water 185
19.3   Materials: Once-Through Systems 198
19.4   Materials: Cooling Tower Systems 203

20.1   Soil Corrosivity to Steel 212

21.1   Atmospheric Contaminants 219

22.1   Nonmetallics in Nitric Acid Service 231

23.1   Materials in HF Environments 249
23.2   Corrosion: Steel in Dilute Sulfuric Acid 252

26.1   Nickel Alloys in Chlorine Service 274

34.1   Economic Calculation Factors 334

Figures

3.1   Stress-strain Diagram 26

3.2   Nil Ductility Transition Temperature 28

3.3   S-N Fatigue Curve 29

3.4   Shielded Metal Arc Welding 35

3.5   Weld Joint Designs 38

3.6   Weld Types 39

3.7   Weld, Fusion Zone (FZ) and Heat-Affected
       Zone (HAZ) 39

3.8   Lack of Penetration 40

3.9   Weld Porosity 40

3.10   Weld Cracking 41

5.1   Polarization 58

5.2   Oxygen Concentration Diagram Cell;
       Lab Demonstration 60

5.3   Copper: Copper Sulfate Reference Electrode ... 61

5.4   Galvanic Corrosion; Water box 66

6.1   Forms of Corrosion, Schematic 71

7.1   Weld Decay 78

10.1   Atomic Structure 106

10.2   Unit Cells 106

11.1   Gray Cast Iron Microstructure 112

11.2   Ductile Cast Iron Microstructure 113

11.3   Iron-Carbon Phase Diagram 115
11.4   Steel Microstructure 117
11.5   Through-Hardening Profile of Quenched Steel 118

12.1   The Gamma Loop 125

19.1   Corrosion of Steel vs Chlorides 187
19.2   Steel: pH vs Corrosion 188
19.3   Steel: Corrosion vs Temperature 188
19.4   Diagram of Boiler 196
19.5   Diagram of Cooling Tower 200
19.6   Water Savings vs Cycles of Concentration 201

27.1   Caustic Cracking Curves: Steel and
       Type 300 Series Stainless Steel 279
27.2   Caustic vs Borosilicate Glass 282

30.1   Simplified Nelson Curves 297

31.1   Time-Deformation (Creep) Curve 302
31.2   Creep Rate and Stress-Rupture Time 303

33.1   Strength-Welded Tube-to-Tubesheet Joint 324

38.1   Galvanic and Impressed CP Systems 373
38.2   Current-Potential Distribution 374
38.3   Low-Resistivity Strata on Current Flow 375
38.4   Foreign Structure 375

Preface to Second Edition
2nd Printing

The first edition of *Corrosion Control in the Chemical Process Industries* (New York, NY: McGraw-Hill, 1986) was very well-received by the by the scientific and engineering communities.

Intended as a primer or elementary text, its purpose was to familiarize the reader, technical and non-technical, with concepts and terminology concerned with corrosion control for materials of construction in chemical processes and ancillary services such as steam generation and cooling water. Basically, it contained no material that was unavailable elsewhere, but centralized much scattered information. It contained technical material of interest also to professional materials/corrosion engineers, although it was addressed *primarily* to their clientele.

Because the material was written over the ten-year period from 1974 to 1984, it was behind the times regarding some materials, processes, and information retrieval. This second edition attempts to address this problem. The author owes much to constructive input from professional associates at the Materials Technology Institute of the Chemical Process Industries, Inc.; to MTI Director A. S. (Bert) Krisher; to fellow consultants of the Nickel Development Institute (NiDI); to friends and associates in various technical societies; and to others too numerous to mention.

I hope this book will be of on-going usefulness to the many technical and non-technical people who are introduced intermittently to the problems of materials selection, corrosion, failure analysis, etc. in the chemical process and other industries.

The technical editing of Dr. Warren I. Pollock at MTI is gratefully acknowledged.

C. P. Dillon
St. Albans, WV

Preface to First Edition

Corrosion control is the real industrial need. The concept of corrosion embraces a broad range of activity from the complete prevention of corrosion through various stages and degrees of protection to simple acceptance of the corrosion situation as it exists, depending upon economic decisions tempered by safety and other considerations.

Troubleshooting and failure analysis of corrosion problems should be secondary to proper materials selection and appropriate corrosion control measures in the design stage. Proper materials selection entails not only elements of design and operation but also provision for corrosion monitoring, inspection, and predictive, or at least preventive, maintenance.

Such efforts must be squarely based on corrosion *technology*, i.e., the reduction to engineering practice of scientific knowledge. Modern advances in materials and corrosion science must be applied at the practical level through adequate interpretation and communication.

The purpose of this book is to introduce the fundamental concepts of corrosion control to working engineers, scientists, and supervisors in the chemical, petrochemical, and other process industries, particularly. The approach chosen discusses fundamental considerations, corrosion phenomena and their mechanisms, the several aspects of corrosion control, attributes of common materials and their corrosion characteristics, and the corrosive nature of common environments.

I am indebted to Professor Ellis Verink of the University of Florida for reviewing much of the text in its first draft and making valuable suggestions for revision, correction, and expansion. I am also indebted to my many colleagues and associates, especially those of Union Carbide Corporation and Aramco, who encouraged me in the preparation and publication of this book.

This book is dedicated to the late Severn M. Frey, Works Chemist at Union Carbide Corp., Texas City, TX, who first introduced me to the fascinating subject of corrosion control in the chemical processes.

Corrosion Control in the Chemical Process Industries

1

Introduction

1.1 Purpose

Corrosion is the deterioration of a material or its properties as a consequence of reaction with its environment, as described in Chapter 4. *Corrosion control* consists of one or more measures (e.g., materials, selection, inhibition, cathodic protection, and special design features) that will diminish the corrosion rate or permit its toleration in specific circumstances. The elements of corrosion control are discussed in detail in Section V.

The primary purpose of corrosion control in the chemical process and related industries is the reliable, continuous manufacture of products for profit. The production and profits depend upon the economical selection of materials of construction and the applicable corrosion control measures, and on the reliability of the manufacturing equipment. These, in turn, depend upon effective communication along a chain of personnel, beginning with the materials/corrosion scientists through various engineering disciplines, purchasing, inspection, construction, operations, and maintenance personnel. It is essential that the people concerned—from management through the scientific and engineering staff down to the craft supervisors—understand at least the fundamentals of materials and corrosion control if a *safe* and productive operation is to be realized.

1.2 Cost of Corrosion

Several studies of the cost of corrosion in developed countries put the figure in the range of three to four percent of their gross domestic product (GDP). In the United States, for example, an estimate of the total cost of corrosion is approximately $80 billion per year. Of this amount, perhaps $55 billion is *irreducible*. The breakdown of materials due to corrosion cannot always be avoided, nor is it always desirable or economical to do so. Well-engineered systems already represent the optimum combination of materials of construction and appropriate corrosion control measures, so that there is no possibility of further economies unless there is a breakthrough in technology. For example, the corrosion of type 316L stainless steel (UNS S31603)[1] equipment during the lifetime of an acetic acid plant simply represents one of the costs of doing business. There may be no more economical material of construction, but the equipment is ultimately destroyed by corrosion.

The remaining $25 billion of the total $80 billion figure is considered to be a *reducible* cost, savings from which could be effected by better application of existing technology. The loss occurs largely, it is thought, due to a lack of effective communication; it represents a summation of engineering mistakes.

Hidden costs which have not yet been fully comprehended and addressed include those arising from excessive use of energy and those arising from the wasteful use of materials which are, or soon will be, in short supply. A nation which has no domestic source of chromium, nickel, or manganese should be on guard against waste of these critical elements by corrosion, mechanical loss, or misapplication.

Environmental considerations also enter into the hidden costs of corrosion. In the real world, one is always seeking a realistic compromise between considerations for the quality of life (e.g., in terms of pollution of the atmosphere or of water sources) and what reasonably can be asked of a manufacturing facility from an economic standpoint.

1.3   Methodology

The basic method of corrosion control is the selection of the proper material(s) of construction and the appropriate means of protection against *unacceptable* rates of attack.

The proper material of construction is the one which is optimum in terms of initial cost, maintenance cost, and durability when evaluated in terms of specific accounting procedures, taxes, depreciation, and the time-value of money. One must accept the ultimate loss of materials as an economic reality when corrosion only can be slowed, and not prevented.

Selection of the proper materials of construction includes considerations such as cost, proper design and fabrication, operation, and maintenance. A material may be corrosion-resistant in its own right or may require one or more corrosion-control measures (e.g., painting, inhibition, electrochemical protection) to prevent premature failure.

Because of human error or unforeseen circumstances, early failures of equipment are always possible. These early failures must be guarded against by proper inspection procedures, priorities, and schedules.

When failures do occur, a failure analysis may be required. The determination of the reason for failure and the exact mechanism is necessary if one is to select a suitable alternative material or take countermeasures against recurrence.

In view of these considerations, this book is organized as follows:

- Section I: *Fundamental factors*—basic considerations, factors in materials selection, and procedures;

- Section II: *Corrosion factors*—corrosion mechanisms, phenomena, and testing;

- Section III: *Materials*—the use of various metals and alloys and the complement of non-metallic materials;

- Section IV: *Corrosive environments*—corrosion characteristics of the natural elements (water, air, and soil), acids, bases and salts, high-temperature exposures, and the appropriate materials for such environments; and

- Section V: *Elements of corrosion control*—the economics of corrosion control; use of materials, linings and coatings; control of the environment; electrochemical techniques; inspection and failure analysis.

To make this book as effective a working tool and guide as possible for the people responsible for the protection of chemical and other process plants or facilities, I have included several reference tables and figures. For the reader who desires a greater depth of understanding, most chapters are accompanied by a list of suggested resource information pertinent to the subject discussed.

1.4   Resources

Many resources are available to support the processes of materials selection and corrosion control.

1.4.1   People

Professional Contacts

Professional contacts are found both in-house and outside the company. These may be professionals (e.g., corrosion and materials engineers, engineering specialists, designers) or simply people who have experience in and a personal concern with the matter at hand (e.g., purchasing agents or buyers, equipment negotiators, operations personnel, maintenance craftsmen, and inspectors). As long as proprietary information is not jeopardized, the same type of people, even in a competitive company, are inclined to share their knowledge in the areas of materials and corrosion.

Consultants

Independent consultants are available, usually for a modest fee in relation to what they have to offer in terms of experience and judgment. Some consultants may be affiliated with a manufacturer, usually because they were associated in some way with development of one or more products (e.g., corrosion-resistant castings). This in no way reflects upon their objectivity, but, rather, indicates that they are only a step removed from the vendor's own technical service department.

Technical Service Departments

The technical service departments of the major producers of corrosion-resistant materials long have been of tremendous help in the resolution of corrosion-related problems. Such departments are the primary source of information on physical and mechanical properties and corrosion characteristics of specific materials.

Unfortunately, there is a growing tendency for manufacturers to dispense with this type of service because it is perceived more as a promotional or public relations effort than one which contributes specifically to "the bottom line." More and more, the burden of collecting and collating technical data has fallen upon the user rather than the individual manufacturer or vendor.

1.4.2   Organizations

Additional sources of information include:

Technical Societies

> NACE International (NACE)
> P.O. Box 218340
> Houston, TX 77218-8340
> Tel: 713/492-0535
> Fax: 713/492-8254

NACE International (formerly the National Association of Corrosion Engineers) is the leading technical society in the fields of corrosion, corrosion prevention and control, and materials selection for corrosive services. It supports a variety of research and educational efforts, and maintains liaison with other organizations with activities in these areas. The technical arm of NACE is the Technical Practices Committee (TPC), comprised of more than 40 group and unit committees concerned with writing standards and exchanging information in specific industries or particular areas of concern. NACE standards consist of recommended practices, materials requirements, test methods, and other documents which reflect the state of the art for a variety of corrosion control and/or materials selection problems.

> American Society for Testing and Materials (ASTM)
> 1916 Race St.
> Philadelphia, PA 19103-1187
> Tel: 215/299-5400
> Fax: 215/977-9679

ASTM is the primary source of specifications relating to corrosion-resistant materials and various kinds of corrosion tests. Various parent committees are concerned with specific types of materials (e.g., A-1 on iron and steel, B-2 on non-ferrous alloys, B-5 on copper and its alloys) while a special committee, G-1, governs a group of subcommittees concerned strictly with corrosion phenomena. It should be noted in passing that specifications have a legal standing not usually accorded to standards.

> Materials Technology Institute of the Chemical Process
> Industries, Inc. (MTI)
> 1215 Fern Ridge Parkway, Suite 116
> St. Louis, MO 63141-4401
> Tel: 314/576-7712
> Fax: 314/576-6078

MTI is a unique, cooperative research organization represent-ing private industry and comprising a number of chemical, petro-chemical, and materials companies. Its goal is to conduct generic, nonproprietary studies of a practical nature on the deterioration of materials and equipment used in the process industries. The end-product is a collection of publications, audio-visual products, and software which is ultimately available to the general public.

> ASM International (ASM)
> Materials Park, OH 44073-0002
> Tel: 216/338-5151
>   800/336-5152
> Fax: 216/338-4634

ASM International (formerly the American Society for Met-als) is concerned with technology in the metallurgical and related fields, including corrosion.

> The American Society of Mechanical Engineers (ASME)
> United Engineering Center
> 345 E. 47th St.
> New York, NY 10017-2392
> Tel: 212/705-7722
> Fax: 212/705-7674

ASME is responsible for administering the Boiler and Pressure Vessel Code, which governs the use of metals and alloys in many industrial applications.

> American Institute of Chemical Engineers (AIChE)
> United Engineering Center
> 345 E. 47th St.

New York, NY 10017-2392
Tel: 212/705-7338
Fax: 212/752-3294

AIChE includes a Materials Engineering and Sciences division concerned with materials and corrosion technology in the chemical process industries.

American Welding Society (AWS)
550 N.W. 42nd Ave.
Miami, FL 33126
Tel: 305/443-9353
Fax: 305/443-7559

AWS has a proprietary interest in corrosion effects related to welding processes.

American National Standards Institute (ANSI)
11 W. 42nd St.
New York, NY 10036
Tel: 212/642-4900
Fax: 212/398-0023

ANSI is the governing organization for many documents relating to materials selection, especially piping.

Any technical society which deals with materials of construction must have at least some interest in the control of deterioration of these materials under the influence of natural or artificial environments.

Trade Associations

Trade associations are organizations formed to promote the proper and profitable utilization of specific types of products and are subsidized by all or most of the major manufacturing companies concerned. Some organizations also concerned with corrosion and materials selection are listed below.

Nickel Development Institute (NiDI)
214 King St., W., Suite 510
Toronto, Ontario M5H 3S6
Canada
Tel: 416/591-7999
Fax: 416/591-7987

American Iron and Steel Institute (AISI)
    1101 17th St., NW, Suite 1300
    Washington, DC 20036-4700
    Tel: 202/452-7100
    Fax: 202/463-6573

Steel Founders' Society of America (SFSA)
    Cast Metals Federation Building
    455 State St.
    Des Plaines, IL 60016-2276
    Tel: 708/299-9160
    Fax: 708/299-3105

In 1970, SFSA absorbed the Alloy Casting Institute (ACI), originator of ACI designations for cast stainless steels.

American Petroleum Institute (API)
    1220 L St., NW
    Washington, DC 20005
    Tel: 202/682-8000
    Fax: 202/682-8030

American Water Works Association (AWWA)
    6666 W. Quincy Ave.
    Denver, CO 80235-3098
    Tel: 303/794-7711
    Fax: 303/794-7310

The Society of the Plastics Industry (SPI)
    1275 K St., NW, Suite 400
    Washington, DC 20005
    Tel: 202/371-1022
    Fax: 202/371-5200

Copper Development Association, Inc. (CDA)
    260 Madison Ave.
    New York, NY 10016
    Tel: 212/251-7200
    Fax: 212/251-7234

Technical Association of the Pulp and Paper Industry, Inc.
    (TAPPI)
    Technology Park

P.O. Box 105113
Atlanta, GA 30348-5113
Tel: 404/446-1400
Fax: 404/446-6947

The Aluminum Association (AA)
900 19th St., NW
Washington, DC 20006
Tel: 202/862-5100
Fax: 202/862-5164

The Chlorine Institute
2001 L St., NW, Suite 506
Washington, DC 20036
Tel: 202/775-2790
Fax: 202/223-7225

Specialty Steel Industry of the United States (SSIUS)
3050 K St., NW, Suite 400
Washington, D.C. 20007
Tel: 202/342-8888
Fax: 202/338-5534

Manufacturers

Although the availability of technical service is diminishing, the manufacturer is likely to be the most reliable source of information on proprietary materials. This is particularly true of manufactured items and composite materials. Further, for many items, such as valves, pumps and compressors, there is a list of manufacturer's standard materials which may limit availability, especially regarding potential alternates.

Historically, the major suppliers of alloys, plastics, and coatings have been repositories for a large amount of both laboratory and field corrosion data and experience. They are reliable sources of information, especially as to where *not* to use their materials.

1.4.3   Literature

Periodicals

Certain journals and magazines are devoted specifically to the scientific aspects (*Corrosion* [NACE International], *Journal of the Electrochemical Society*) or the technology (*Materials Performance* [NACE International], *Advanced Materials and Processes, Welding Journal*) of corrosion and materials. Others have peripheral interests evident in particular articles or specific sections (*Chemical Engineering Progress, Petroleum Refiner,* and *Hydrocarbon Processing*). Trade journals like *Plant Engineering* and *Maintenance Engineering* are also often useful sources of information. *Nickel,* published by Nickel Development Institute (NiDI), has case histories on the uses of nickel-containing alloys.

Abstracts

Professional abstracts services offer publications which summarize both domestic and foreign articles on corrosion and related matters. The two major publications are *Corrosion Abstracts* (NACE International) and *Corrosion Prevention and Inhibition Digest* (ASM International).

Manufacturers' Literature

An excellent starting point for many corrosion/materials investigations is the manufacturers' literature. Most major manufacturers offer brochures which summarize the properties, including corrosion resistance, of their products. Among the more sophisticated presentations available are "Corrosion Engineering Bulletin" published by the International Nickel Company (Inco) and now available from NiDI, as well as analogous information available from Avesta Sheffield, Haynes International, Rolled Alloys, Teledyne Wah Chang, VDM Technologies (Krupp VDM), and many others, especially information concerning their proprietary products. For products produced by many companies (e.g., the type 300 series austenitic stainless steels), there are a large number of summaries available.

Reference

1. *Metals & Alloys in the Unified Numbering System,* (Warrendale, PA: SAE).

Suggested Resource Information

ASM Handbook, Vol. 13, *Corrosion* (Materials Park, OH: ASM International, 1987).

J. T. N. Atkinson, H. Van Droffelaar, *Corrosion and Its Control–An Introduction to the Subject* (Houston, TX: NACE International, 1984).

B. D. Craig, ed., *Handbook of Corrosion Data* (Materials Park, OH: ASM International, 1989).

C. P. Dillon, *Materials Selection for the Chemical Process Industries* (New York, NY: McGraw-Hill, 1992).

M. G. Fontana, *Corrosion Engineering*, 3rd Ed. (New York, NY: McGraw-Hill, 1986).

P. J. Gellings, *Introduction to Corrosion Prevention and Control* (Delft, The Netherlands: Delft University Press, 1985).

R. J. Landrum, *Fundamentals of Designing for Corrosion Control–A Corrosion Aid for the Designer* (Houston, TX: NACE International, 1989).

B. J. Moniz, W. I. Pollock, eds., *Process Industries Corrosion– The Theory and Practice* (Houston, TX: NACE International, 1987).

P. A. Schweitzer, *Corrosion Resistant Tables*, 3rd Ed. (New York, NY: Marcel Dekker, 1991).

R. S. Treseder, R. Baboian, C.G. Munger, eds., *NACE Corrosion Engineer's Handbook*, 2nd Ed. (Houston, TX: NACE International, 1991).

H. H. Uhlig, R. W. Revie, *Corrosion and Corrosion Control*, 3rd Ed. (New York, NY: Wiley-Interscience, 1985).

L. S. VanDelinder, *Corrosion Basics–An Introduction* (Houston, TX: NACE International, 1984).

SECTION I

Fundamental Factors

2

Basic Considerations

The five basic considerations in corrosion control for the chemical process industries, in descending order of importance, are: safety and reliability, cost, environmental factors, energy considerations, and materials conservation.

2.1   Safety and Reliability

Safety and reliability are of primary importance. One must consider not only the immediate safety of the equipment as it relates to profitable operation, but also the short- and long-term effects upon plant personnel and even upon the local community.

The selected corrosion control measures must be considered relative to:

● Fire hazards;

● Explosion hazards;

● Brittle failures;

● Mechanical failures; and

● Release of toxic, noxious, or other hazardous materials, except with adequate safeguards (e.g., scrubbing systems).

Fire and explosion are the most dramatic incidents among corrosion failures. One should be aware of the hazards inherent not only in the leakage of process fluids because of corrosion failures, but also of the nature of some of the corrosion products themselves. Some of the problems to keep in mind are:

● Pyrophoric iron sulfides and certain sulfur-based corrosion products of zirconium become red-hot on exposure to air or oxygen;

- Explosive compounds are formed by reaction of some corrosion products (e.g., those of copper, silver, or mercury) with acetylene;

- Explosive silver azides are formed from reaction of ammonia or amines with silver salts;

- Certain organic compounds (e.g., diacetyl peroxide) become explosive on drying; and

- Any organic materials can react explosively with powerful oxidizing agents, such as perchloric acid.

Be aware of corrosive situations that are made worse by poor control of process variables, actual or potential. Some things to watch out for include:

- Unanticipated temperature excursions within the process equipment. Added to normal corrosion, such excursions can be disastrous. Temperatures may rise because of the failure of a cooling apparatus, or from exotherms occasioned by contamination (as from the inadvertent ingress of alkaline substances into acrylic esters);

- Contamination of a process stream, such as water leaking into a chlorinated hydrocarbon or chloride contamination of an organic acid, may cause rapid unexpected attack; and

- Recycling a process stream previously discharged to sewer (e.g., introducing ferric ions, increasing chloride levels).

Other items which the corrosion engineer should be aware of (items that might seem unrelated to corrosion inspection, control, and remedy) include:

- Unscheduled shutdowns which sometimes may cause corrosion, especially if they preclude special protective measures prescribed for normal operations to neutralize corrosive species (e.g., alkaline washes for stainless steel equipment in polythionate service); and

- Hazardous materials, themselves noncorrosive, which might be released to the atmosphere. It is important to realize that some materials become corrosive upon exposure to atmospheric moisture (hydrogen chloride, chlorine). Poisons (hydrogen sulfide, phosgene, methyl isocyanate) or carcinogens such as vinyl chloride monomer (VCM) and polychlorinated biphenyls (PCBs) must be contained.

It is the duty of the corrosion engineer to anticipate and prevent corrosion failures related particularly to such potential dangers.

Examples of some specific hazards and related preventive measures are given in Table 2.1 (and in other sections relative to specific materials and environments).

Misapplications of conventional materials also may be hazardous, for example, steel drain plugs in alloy pumps; use of the wrong welding rod (type 347 [UNS[1] S34700] consumable in type 316L [S31603] equipment); failure to blank off nonresistant items during chemical cleaning (e.g., stainless steel-trimmed valves from inhibited hydrochloric acid, copper alloys from ammoniated citric acid, aluminum components from alchohol, or chlorinated solvents).

It should be evident that careful selection, design, and operation are of the utmost imprtance to ensure safety and reliability and that equipment must be inspected at regular intervals. Inspection techniques are discussed in detail in Chapter 39.

2.2   Cost

In seeking the optimum materials of construction (see also Chapters 3 and 34)—in whatever combinations of initial cost, corrosion resistance, amenability to corrosion control, and service life—the selection is inevitably compromised to some extent. It is not economical to design for unlimited life. Management will require a financial justification that will include not only the first cost of materials and labor, but also the cost of downtime and the aspects of safety and environmental control. Unforeseen or uncontrolled hazardous failures must be precluded, which, not only entails proper materials selection and design, but also predicative

---

[1] See Chapter 3.4.5., Unified Numbering System.

Table 2.1

Specific Hazards and Related Protective Measures

Hazard	Remedy
Localized Failures	
External corrosion of steel under wet insulation	Heavy-duty industrial coating under insulation
External corrosion of steel chlorine piping at pipe supports	Pipe slides; plastic sheathing; heavy-duty coatings
Localized corrosion of underground gas lines	Coatings plus cathodic protection
Erosion-corrosion of steel sulfuric acid piping	Reduced velocity; long-radius ells; substitute materials
Special Effects in Corrosion	
Selective dissolution of alloy constituents	Proper alloy selection
Specific hydrogen phenomena	Alloy selection; environmental control
Hot-wall effects in heat exchangers	Alloy selection; process changes
Cracking Phenomena	
Mechanical	
Brittle fracture (NDT of steels at low temperature, superferritic stainless steels)	Limit thickness; minimize flaws and notches
Fatigue applied stress; shot-peen; auto-frettage	Avoid sharp discontinuities; lower
Corrosion-Related	
Stress Corrosion Cracking (SCC)	Materials selection; stress relief
Hydrogen-Assisted Cracking (HAC)	Materials selection; hardness control
Liquid Metal Cracking (LMC)	Materials selection
Corrosion Fatigue	Avoid sharp discontinuities (as for fatigue); materials selection

(or at least preventive) maintenance, based upon adequate monitoring and inspection procedures.

Cost estimates must be based on total (not first) costs calculated on an annual or life-cycle basis.

2.3   Environmental Aspects

Concern for the quality of air and effluent industrial water extends far beyond the immediate concerns relative to safety and toxicity, as discussed in Chapter 2.1. Atmospheric pollutants include corrosive species (e.g., sulfur dioxide, oxides of nitrogen, hydrogen sulfide, hydrogen chloride, etc.) as well as those which can be either autocorrosive or have adverse catalytic effects upon other contaminants (e.g., coal dust). Many organic vapors contribute to overall pollution, increasing health hazards in terms of eye irritation or pulmonary problems.

Effluent water must not only be free of biocides, carcinogens, and other objectionable species, particularly in streams which might contaminate a municipal water supply, but even "thermal pollution" may be objectionable. Thermal pollution occurs when a cooling water discharge raises the ambient temperature of the receiving body of water with possible adverse effects on commercial fishing, for example.

In developing countries, it is obvious that some compromise must be effected between the need for industrial growth and concern for environmental quality. However, care must be taken that such pollution as can be tolerated temporarily does not cause irreversible damage.

It is unfortunately true that many industries in developed countries grew to considerable size without adequate appreciation of the damage they were inflicting on the environment. Originally, such damage was seen by all concerned as a necessary price to pay for industrial growth and job opportunities. In today's climate, such industries must be prepared to clean up their atmospheric and aqueous effluents with all deliberate speed and with adequate consideration of the economic demands entailed. Most reputable firms today willingly embrace even multimillion-dollar environmental control programs, once the need becomes apparent, if they are permitted to do so at a rate that will not destroy them in a competitive market.

A substantial amount of the corrosion and materials engineer's technical input into a project may be concerned with its possible environmental impact, and this in turn, may have a substantial influence on the economics of materials selection, as discussed in Chapter 4.

2.4   Energy Considerations

Industry has always had to consider the availability of fuel and cooling water in selecting plant locations, in addition to the basic considerations of raw materials and transportation. It takes 10 to 15 tons of water to produce one ton of gasoline or sulfuric acid; several hundred tons for each ton of ammonia, steel, paper pulp or rayon; and, several thousand tons per ton of aluminum.

The quality and amenability to treatment of fuel and water resources has not always received proper consideration. Fuel quality considerations include inherent corrosion problems such as vanadium contamination in oil stocks, and indirect environmental effects such as sulfur dioxide emissions from high-sulfur coal, one of the causes of acid rain.

Water must be available in sufficient quantity and either be of an inherently suitable quality or be amenable to economic treatment against scale and corrosion, as for cooling purposes or steam generation.

The cost of energy for a proposed plant or process must be accurately appraised, including steam losses. The mnemonic "TILT" often is used to remind one to consider *t*raps, *i*nsulation, *l*eaks, and *t*racing—the major sources of wastage of steam.

2.5   Materials Conservation

In some time periods, certain elements may be in critical supply (especially in the United States); for example, chromium, nickel, and manganese.

Although such shortages may reflect only political circumstances, these are essential elements in the manufacture of corrosion-resistant and high-temperature alloys, and consideration must then be given to their possible recovery and reuse. This will not always be possible, but because the high alloys usually are used where they are in fact suitably resistant, much of the domestic usage can be recovered by adequate salvage and recycling plans.

In the past, the free enterprise system has promoted the entrepreneurial manufacture and sale of products generally without regard to the long-term needs of the economy. With present-day knowledge of the finite nature of resources, elements in critical supply should be discouraged from use unless they contribute to the real service life. For example, stainless steel in many cosmetic applications (e.g., trim on kitchen or other appliances) can be replaced with other suitable materials of equally aesthetic appearance.

3

Factors in Materials Selection

The major factors to be considered in materials selection are cost, physical and mechanical properties, availability, fabricability, corrosion characteristics, and amenability to corrosion control. Although the reader with an engineering background already will be familiar with the properties of engineering materials, the following discussion will aid in recalling their relationship to corrosion problems.

3.1   Cost

Although the initial cost of materials has an important psychological effect, it is essential that the *true* (rather than first) cost be considered. The true cost of process equipment includes materials, labor of fabrication and installation, annual maintenance, salvage value, and cost of corrosion control, if any. In comparing alternatives, a common basis is their *equivalent uniform annual cost*, which permits direct comparison of different service lives and different categories of expenditure. For alternatives of identical life, *life-cycle costs* may be employed. These concepts are explained in Chapter 33, on the economics of corrosion control.

3.2   Physical Properties

Some physical properties are more important than others, and they are generally less important than mechanical properties from the engineering standpoint. However, they should be taken into consideration as required because they do have some influence on corrosion in some instances.

3.2.1   Specific Gravity

Specific gravity is the ratio of mass per unit volume to that of water, whose specific gravity equals 1 g/cm^3. It is a factor in the strength:weight ratio, in ease of physical handling of equipment, and in coverage per unit weight by metallic coatings. For example, a given weight of aluminum applied to a given area of steel substrate will give a thicker coating than will the same weight of zinc (approximately 2.5 times thicker).

However, zinc is preferred as a sacrificial coating (e.g., hot-dipped galvanizing), whereas aluminum coating is preferred as a barrier for high-temperature applications (e.g., "Calorizing").

3.2.2   Thermal Conductivity

Thermal conductivity is a measure of the capacity of a material to conduct or transfer heat. It is of importance primarily in heat exchanger design. Thermal conductivity is greater in pure metals than in their alloys. However, from an engineering standpoint, the thermal conductivity of the metal per se is often overridden by surface film effects in practical applications. Because a metal transferring heat to an environment will sometimes corrode at a higher rate than the same metal simply immersed at equivalent temperature, it is sometimes necessary to use different materials for calandria (reboiler) tubes or heating coils than for the vessel proper (e.g., silver coils in type 316L acetic acid tanks or type 316L coils in aluminum tank cars for monoethanolamine).

3.2.3   Thermal Expansion

Thermal expansion is expressed as a decimal part of an inch per inch (in millimeter per millimeter) per unit of temperature (e.g., 11.7 millionths of an inch/inch/degree Celsius, as for pure iron at 20°C [68°F]). The practical significance of thermal expansion lies in the stresses caused by *differences* in expansion between different components of equipment (e.g., tubes vs shell in a heat exchanger), and in the necessity of making provision for the elongation or contraction of engineering structures.

Thermal expansion problems usually are associated with improper design but can have an inherent influence on corrosion problems. For example, it is impractical to thermally stress-relieve steel vessels clad with austenitic stainless steels because the difference in thermal expansion will still leave the cladding in tension and potentially susceptible to stress corrosion cracking (SCC).

3.2.4   Melting Point or Range

Pure elements have melting *points*, but most engineering materials have a melting range rather than a distinct melting point. The practical significance of this relates largely to weldability (e.g., a material with a *narrow* melting range is more difficult to weld with a structurally sound weldment). "Hot-short cracking" is brittleness in metals, due to the presence of low-melting constituents at the grain boundaries during solidification. Hot-short cracking can be a problem in hot-forming as well as welding (e.g., in type 347 [UNS S34700] and in high silicon bronze A (C65500). Hot-short cracks may in turn become focal points for the inception of corrosive attack (e.g., SCC, corrosion fatigue, crevice corrosion).

3.2.5   Magnetic Properties

Magnetic properties are rarely an engineering concern, except as they relate to identification, methods of inspection, or removal of particulate matter, such as "tramp" iron or mill-scale, from a product stream.

3.2.6   Modulus of Elasticity

Although a physical property, the *modulus of elasticity* (Young's modulus) is the ratio of stress to strain below the yield point. The unit is pressure or stress (force per unit area)/unit length/unit length, namely, psi/inch/inch or psi (pascal [Pa]). The modulus of elasticity for many steels and ferrous alloys is approximately 30 million psi (210 million kPa).

3.3   Mechanical Properties

Mechanical properties relate to the strength and physical performance of process equipment and ancillary structures.

3.3.1   Tensile Strength

The ultimate, or tensile strength, of a material is a primary engineering concern. It is a measure of the load a material can sustain more or less indefinitely. This property can be changed radically by cold-work, heat treatment, or both, depending upon the nature of the material.

3.3.2 Yield Strength

A few materials have a definite yield point, at which there is a temporary drop in stress with an increase in strain. Most engineering alloys are characterized by an arbitrary *yield strength* (e.g., at 0.02% offset), above which they will no longer behave in an elastic manner. A value of five-eighths of the yield strength is often used as the maximum allowable applied stress for mechanical design.

The yield point, tensile strength, and modulus of elasticity are illustrated in Figure 3.1, a stress/strain diagram for a hypothetical alloy. In the process of developing the data for such a diagram, one also can obtain a measure of ductility, as noted below.

3.3.3 Elongation and Reduction in Area

Elongation and reduction in area are complementary measures of the ductility of a material. *Elongation* is expressed as the amount a material will stretch, due to applied stress over some defined unit length, before breaking under a tensile load. *Reduction in area* is the corresponding diminution in cross-sectional area at the fracture. Greater elongations are associated with greater reduction in area (i.e., "necking down"). Conversely, the smaller the value for each, the less ductility the material has exhibited.

Ductility (and toughness, see Chapter 3.3.4) is an important consideration in mechanical design, especially under conditions of tensile loading.

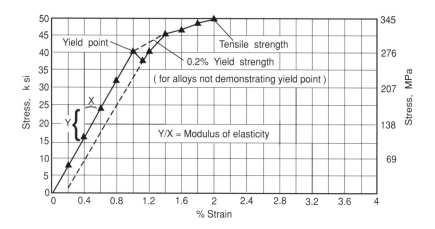

Figure 3.1 Stress-strain diagram.

Notch sensitivity is an important auxiliary consideration. An otherwise ductile material may nevertheless break in a brittle fashion if it is notch-sensitive. For example, threaded pipe connections in a "rimmed" steel may break off rather easily, even though the parent metal of the pipe has good ductility in the unnotched condition.

3.3.4 Toughness

Toughness is the ability to withstand impact, and is related in a general way to ductility. There are, however, many engineering materials which lose their toughness with decreasing ambient temperature to a degree that is determined by *thickness*. For example, ordinary carbon and low-alloy steels, as well as ferritic stainless steels, which are usually ductile, can fail under impact in a brittle manner as temperatures are lowered even slightly below room temperature, particularly as thickness is increased. This dependency on thickness indicates that toughness under impact is not a true physical or mechanical property. It is discussed under mechanical properties because it is an important element in mechanical design.

Toughness is evaluated primarily by means of high-impact tests, using standardized specimens (e.g., Charpy V-Notch), which measure the absorption of energy of a given thickness of material at a given temperature. The values at different temperatures are then delineated in a Nil Ductility Transition Temperature (NDTT) diagram (Figure 3.2). Other types of tests (e.g., drop-weight, tear tests) have also been devised to study this phenomenon.

Selection of some materials, such as superferritic stainless steels, for their outstanding corrosion resistance in some applications may be limited to thin-walled components (e.g., heat-exchanger tubes) because of the danger posed by high NDTT for heavier-walled components, such as pipe or vessel plate.

3.3.5 Hardness

Hardness, or resistance to indentation, is related in a general way to strength and resistance to erosion-corrosion or abrasion. Hardness may be increased by cold-work, heat-treatment, or both, depending upon the nature of the material. It is expressed in

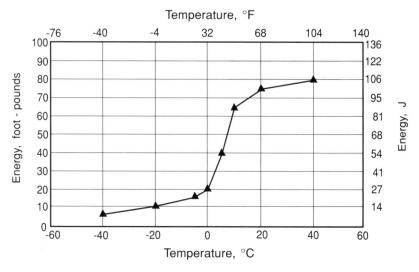

Figure 3.2 Nil ductility transition temperature diagram.

arbitrary units (e.g., Brinell hardness number [HBN]; Rockwell B or C [HRB or HRC]), depending upon the type of indenter employed and the static applied load used in the particular hardness test.

Hardness may cause corrosion problems where there is a specific relationship between internal stresses and corrosion phenomena, as in sulfide stress cracking (SSC), discussed in Chapter 8.1.

3.3.6 Fatigue Strength

Fatigue strength is the ability of a material to withstand repeated alternations of stress at some specific stress level. An "S-N" curve can be derived which describes the number of *cycles*, i.e., repetitions, of alternating stress which can be withstood at specific levels of stress (Figure 3.3).

In a general way, fatigue strength increases with increasing yield strength (and hardness, which has a rough correlation with yield strength), at least until notch sensitivity develops to counter-act the benefit of higher strength. For example, a chromium-molybdenum 4130 steel (UNS G41300) has optimum fatigue strength at a hardness of approximately HRC 35, above which value it is harder and stronger but less reliable in a situation of cyclic stress (due to notch sensitivity).

The *endurance limit* is that stress level at which (or below which) the material will endure cyclic stress indefinitely. However, in the simultaneous presence of corrosive action, there is *no* endurance limit. In cases of *corrosion fatigue*, the S-N curve continues to drop with increasing cycles of repetition; no endurance limit is to be found.

3.4   Codes and Regulations

Two specific codes and a number of standards and specifications are of interest.

3.4.1   ASME Boiler and Pressure Vessel (B&PV) Code

Historical incidents of pressure vessel failure, notably of steam boilers, led to the development of regulatory codes relating to design and inspection in the public interest under the aegis of the American Society of Mechanical Engineers (ASME) Boiler and Pressure Vessel Committee.

The committee's function is to establish rules of *safety* governing the design, fabrication, and inspection during construction of boilers and pressure vessels, and interpret these rules when questions arise regarding their intent. In formulating rules, the committee considers the needs of users, manufacturers, and inspectors of pressure vessels. The objective of the rules is to afford reasonably certain protection of life and property, and to give a reasonably

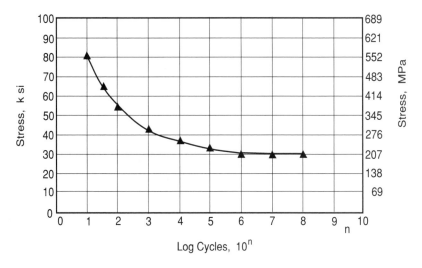

Figure 3.3 S-N fatigue curve.

long and safe period of usefulness. Advancements in design and materials, and the evidence of experience have been recognized.

The ASME B&PV Code[1] applies to both fired (Section I) and unfired (Section VIII) pressure vessels, using Section II-approved specifications only, with the exception of permitting code cases to develop experience. Even then, those materials must be listed in a specification approved by Section II before the material is allowed. Section II contains ASME-accepted materials specifications: Part A, Ferrous Materials; Part B, Nonferrous Materials; Part C, Welding Rods, Electrodes, and Filler Metals. A new Part D, Properties, was issued in May 1992, and includes allowable stress values, in addition to physical properties, for all materials in the ASME Code. (These values were formerly listed in the respective book sections [I, II, VIII] in the 1989 edition of the Code.)

Not all ASTM materials specifications are included in ASME; the Code contains only those with proper identification of material, chemistry, properties, working, and testing. Because ASME is only concerned with pressure, many ASTM specifications never have been requested to be included in ASME.

New materials and designs must receive proper approval on a Code Case basis before being used as ASME-approved materials in a plant. Analogous codes have been developed in other countries.

3.4.2   ASME Code for Pressure Piping

A similar code has been developed for pressure piping, B31, the ASME Code for Pressure Piping, an American National Standards Institute (ANSI) standard. The several sections include:

- B31.1   Power Piping
- B31.3   Chemical Plant and Petroleum Refinery Piping
- B31.4   Liquid Transportation Systems for Hydrocarbons, Liquid Petroleum Gas, Anhydrous, Ammonia, and Alcohols
- B31.5   Refrigeration Piping
- B31.8   Gas Transmission and Distribution Piping
- B31.11   Slurry Transportation Piping Systems

The B31 committee has published ANSI/ASME B31G, titled "Manual for Determining the Remaining Strength of Corroded Pipelines," which is a supplement on corrosion to the B31 Code for Pressure Piping.

3.4.3   U.S. Department of Transportation

The U.S. Department of Transportation (DOT) "Hazardous Materials Regulations" governs the shipment of potentially dangerous chemicals.

3.4.4   Specifications and Standards

As discussed in Chapter 1, specifications are documents legally prescribing certain requirements as to composition, mode of manufacture, physical and mechanical properties, freedom from defects, etc. The ASME and ANSI codes refer in detail to certain ASTM product specifications.

Although standards lack the legal aura of specifications, they are accepted both nationally (ANSI) and internationally (the International Organization for Standardization [ISO]) as documents embodying a voluntary consensus. In representing the "state of the art" as to what is a good and reliable way to achieve a specific end result, both standards and specifications are invaluable in engineering practice. In addition to the NACE standards, other standards are available from technical organizations such as the American Petroleum Institute (API), and Society of Automotive Engineers (SAE).

Individual companies also develop engineering standards which reflect their experience and needs. The impact of various codes, standards, and regulations on plant layout have been discussed.[2]

3.4.5   Unified Numbering System

Many trade organizations formerly had their own system for codifying materials (e.g., SAE had a numbering system for steels; AISI for stainless steels; CDA for copper alloys; AA for aluminum alloys). These have been adapted to and replaced by the Unified Numbering System (UNS) developed jointly by ASTM and SAE.[3] The UNS system, for the most part, incorporates the older numbering codes into its system. For example, type 304 is UNS S30400; copper alloy CDA 706 is UNS C70600; and nickel alloy 200 is UNS N02200. In the long run, the UNS should replace the older designations, company names, and historical names which were often misleading. Use of proprietary trade names can be avoided by using UNS designations, e.g., S32654 for alloy 654 SMO[†], N08367 for alloy AL-6XN[†], and N10675 for alloy B-3[†]. (See also Chapters 10.3.1 and 11.3.)

3.4.6 Foreign Designations

There are also foreign designations which may be encountered, among them DIN numbers (German), SIS (Sweden), "Euronorm" (European Community), SUS (Japan), GOST (Russian), and other national systems. ISOs are international standards.

3.5 Fabrication Characteristics

The types of products available in specific alloys and their response to routine working and joining operations are important considerations.

3.5.1 Available Products

The practical usefulness of materials of construction may be limited by available form. Some metals or alloys are not available as castings (e.g., tantalum), whereas other materials, such as heat exchanger tubes, are available only in cast form (e.g., silicon cast irons) and cannot be purchased.

Following is a brief discussion of the more common forms in which metals and alloys are sold.

Castings

A casting is a product made by pouring molten metal into a mold. Pumps, valves, and furnace tubing are examples of products commonly manufactured in this manner.

Forgings

A forging is a cast ingot that has been hammered or pressed into shape under intense heat. Forgings are used for aircraft landing gear, high-pressure parts, hubbed tubesheets, and other massive components in which a more homogeneous structure is needed than can be obtained by casting alone.

[†] Trade name

Wrought Products

In one context, this term is applied to the final product derived from *working* (i.e., rolling, drawing, or extruding) an ingot to a relatively massive final product, such as bar stock. The term *wrought* is an Old English form of the past participle of the verb, "to work." We also use the term "wrought" to denote the small, relatively equiaxed microstructures resulting from such a working; so, the term also applies to flat products.

Flat Products

Flat products include plate, sheet, and strip, which are used either as materials for the fabrication or lining of tanks and vessels, or as a form from which other products (e.g., clad plate, welded pipe, and welded tubing) are manufactured.

Skelp is a flat product destined for the manufacture of pipe or tubing by rolling and forming it longitudinally and welding the edges together, with or without filler metal.

3.5.2 Heat Treatment

Metals and alloys are subjected to heat treatment for a variety of reasons. Usually, the heat treatment is intended to control metallurgical, mechanical, or corrosion characteristics or to effect thermal stress relief. Some of these treatments are discussed in detail in the chapters on materials and phenomena.

Metallurgical characteristics of steels are discussed in more detail in Chapter 11. For the moment, it can be noted that *normalizing* assures homogeneity of structure. The *grain size* of metals and alloys may be controlled for specific purposes, such as to obtain improved low-temperature properties.

The complex interplay between strength and ductility in many steels is controlled almost entirely by the sequence of hardening and subsequent drawing or tempering to effect a suitable compromise between hardness and ductility or toughness.

Thermal stress relief is commonly employed either for mechanical reasons (e.g., to minimize internal residual triaxial stresses), or to improve resistance to environmental cracking or corrosion fatigue.

Homogenization and other effects associated with heat treatment may have a profound influence on corrosion resistance (e.g., of aluminum alloys in certain atmospheric exposures; of certain austenitic stainless steels in environments conducive to environmental cracking; and of corrosion-resistant castings in aggressive environments).

3.5.3 Forming Characteristics

Forming characteristics are of primary concern to the manufacturer. The corrosion/materials engineer, however, should at least be aware of certain problems.

Hot shortness, for example, is a problem both in manufacture or forming and in welding operations. Hot shortness is a cracking tendency encountered when mechanical stress is applied in a critical elevated temperature range due to a low-melting constituent at the grain boundaries or a phase having reduced strength at elevated temperatures. Silicon bronzes, nickel-molybdenum alloys, stabilized austenitic stainless steels, and "duplex" (i.e., mixed austenite-ferrite structure) stainless steels are notorious in this regard.

This undesirable characteristic not only limits hot-formability, as in forming dished heads, but also shows up as cracking during or after welding under conditions of restraint.

3.5.4 Welding or Joining

Brazing, also called soldering at temperatures below approximately 425°C (800°F), uses low-melting alloys to join the metal parts without actually melting the parent metal. Although these are nonstandard terms, according to the American Welding Society, solders are described in common parlance as *soft* or *hard*. Lead- and tin-based alloys comprise the *soft solders*, which have a liquidus below 450°C (850°F). *Brazing materials* (formerly called Hard Solders) are silver-based alloys with copper and nickel additions and a liquidus above 450°C (850°F) but below the solidus of the base metal. When used to join nonferrous metals above approximately 425°C (800°F) and below the melting range of the parent metal, the process is sometimes called brazing.

Welding

In a general way, the term *welding* includes any method of joining metals by the application of heat. In its original connotation, it included pressuring or hammering the metal to facilitate the joining and induce recrystallization across the joint. In current terminology, the concepts of soldering (e.g., *tinkering*) and brazing are excluded; welding indicates the pieces are heated until molten and fused at the joining interfaces of the parent metal.

Pressure welding, which derives from the blacksmith trade, may still be employed for specific products, such as butt-welded pipe. There are still some complicated automatic butt-welding processes used in manufacturing. However, most of the welding which will be encountered by the engineer in the shop or in the field will be one of the processes discussed below.

    (1)   *Oxyfuel Gas Welding*. In this process, a heating torch (e.g., oxy-acetylene) provides the heat for melting both the appropriate filler metal and the edges to be joined. Except for repair welding of some high-temperature castings, this process is rarely used today in industrial plants. Furthermore, it is completely unsuitable for most corrosion-resistant alloys.

    (2)   *Shielded Metal Arc Welding* (SMAW). Stick welding, as it is often called, employs a wire core covered with a special electrode covering (Figure 3.4). Coalescence is effected by an electric arc between the covered electrode and the work. The hot, molten weld deposit is

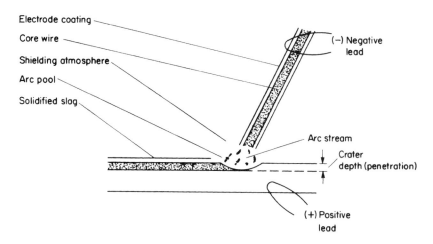

Figure 3.4 Shielded metal arc welding–straight polarity.

shielded from the atmospheric oxygen by the decomposition of the flux coating while the electrode core wire provides the filler metal.

Either alternating current (AC) or direct current (DC) may be employed in this as in other electric-arc welding techniques. In DC welding with straight polarity, as shown in Figure 3.4, the electrode is negative, and the work, positive; current flows from the work through the arc to the electrode. In reverse polarity, the current flow is in the opposite direction and can lead to stray current corrosion (often called *electrolysis*) if an adequately sized return to the DC source is not provided (see Chapter 5.2). Most stainless steel, nickel-based alloys, and low-hydrogen electrodes are used with DC electrode positive (DCEP, or reverse polarity) techniques.

(3) *Gas Metal Arc Welding* (GMAW). This process, previously called MIG, uses a bare consumable metal electrode, but replaces the flux coating with a *shield* of inert gas flow (e.g., helium, argon, carbon dioxide). The gas composition must be carefully selected and controlled in order for the alloy composition to be welded. For example, a shielding gas with more than 2% carbon dioxide is not suitable as a shield for a low-carbon stainless steel to be used in corrosive service, because of the probable carbon pick-up by the molten metal. Reverse polarity is routine for this type of welding, and stray current corrosion problems have been encountered in the welding of buried stainless steel piping, for example. *Short-arc welding* is a special type of GMAW, using a very fine wire and depositing weld metal by short-circuit transfer.

(4) *Submerged Arc* (SAW). This process uses a bare consumable wire electrode.

However, the arc is struck through, and the molten weldment protected by, a pool of molten flux.

(5) *Gas Tungsten Arc* (GTAW). This process, previously called TIG, employs an inert gas shield (as with GMAW) but around a nonconsumable tungsten electrode. The weld is made by melting the base metal either in an autogenous weld (i.e., without filler metal) or by adding a bare filler metal.

Filler Metals

There are many different filler metals designed to meet specific needs. It is important to remember that the filler metal (and its covering, if any) is designed to deposit an appropriate weld metal composition. This is usually, but not always, a close match to the parent metal. Also, alloying elements are sometimes introduced via the covering; a chemical analysis of the *wire*, stripped of its flux covering, is not necessarily representative of the weld deposit.

Examples of weld rods *not* identical with the parent metal to be joined are as follows:

(1) Type 347 rod (Fe-18%Cr-8%Ni+Cb) is used to weld type 321 (Fe-18%Cr-8%Ni+Ti) stainless steel because titanium does not transfer well across the arc. A small amount of ferrite (2 to 5%) in the weld is required to combat hot-short cracking;

(2) Type 308 rod (Fe-19%Cr-9%Ni) is used in welding type 304 stainless steel to enrich the alloy content of the weld and minimize depletion by oxidation effects;

(3) Type 309 (Fe-25%Cr-12%Ni) or type 310 (Fe-25%Cr-20%Ni) rod is used to weld carbon steel to type 300 series stainless steel, to ensure an alloy-rich weldment of suitable ductility; and

(4) Alloy 600-type (Ni-16%Cr-8%Fe [UNS N06600]) rods of high nickel content are used

for a number of dissimilar metal joints to ensure ductility, impact properties, and corrosion resistance.

Welding Positions

The orientation of the proposed weld affects the choices of certain variables because of the influence of gravity on the molten weld metal. Down-hand (flat) position makes the weld made from the upper side of near-horizontal work. In the *vertical* position, the axis of the weld is upright. In the overhead position, one is welding from the bottom side of horizontal work (e.g., from inside the roof of a tank). *Penetration* (i.e., the depth of the fusion zone beneath the surface of the work) may be reduced in overhead welding.

Joints and Welds

Five types of joint design (butt, corner, lap, edge, and tee) are illustrated in Figure 3.5, and four types of welds (fillet, groove, bead, and plug) are shown in Figure 3.6. The anatomy of a grooved butt-weld is shown in Figure 3.7 to indicate especially the fusion

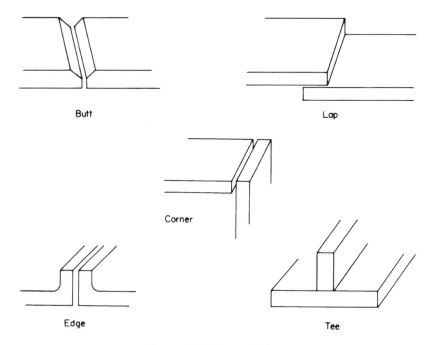

Figure 3.5 Types of joints.

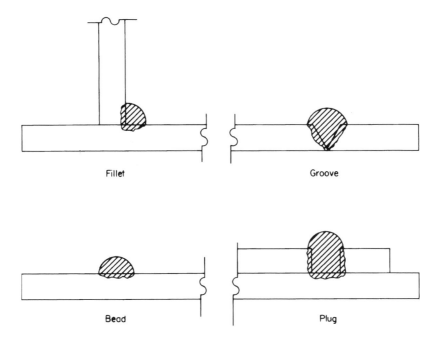

Figure 3.6 Types of welds.

zone and the sensitized weld heat-affected zone (HAZ) as it occurs, for example, in type 300 series stainless steels. The latter, particularly, is often of great significance from a corrosion stand-point in a number of metals and alloys.

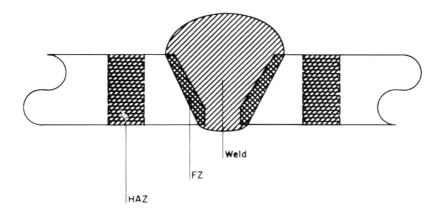

Figure 3.7 Weld, fusion zone (FZ), and heat-affected zone (HAZ).

Weld Defects

There are several types of weld defects (or faults) that are of concern mainly because of mechanical integrity, such as strength, warping, and brittle behavior.

The contours and finished surface have not only aesthetic importance, but also affect the application of paints or coatings for corrosion protection. Several other types of defects may affect not only mechanical integrity but also the initiation or promotion of corrosion phenomena. These include the following:

(1) *Lack of Penetration* (Figure 3.8)—a focal point for accumulation of corrosive species (e.g., of caustic in steam lines) or a site to initiate concentration cell corrosion, environmental cracking, or corrosion fatigue;

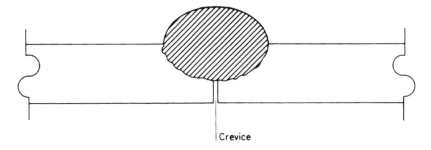

Figure 3.8 Lack of penetration.

(2) *Porosity* (Figure 3.9)—potential site(s) for pitting, cracking, and other corrosion phenomena;

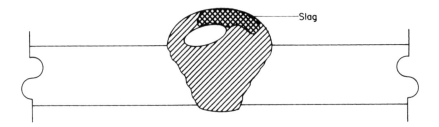

Figure 3.9 Weld porosity.

(3) *Slag Inclusion*—residues of welding flux in the form of slag may corrode out in certain environments, leaving sites with the same potential for problems as derived from weld porosity; and

(4) *Weld Cracking* (Figure 3.10)—this may initiate fatigue, brittle failure, environmental cracking, or act as a site for initiation of corrosion phenomena.

The stresses left in weldments are not defects because they are an inherent part of the cooling of the molten weldment under restraint, but they are conducive to cracking phenomena. They should be reduced by appropriate thermal or mechanical stress relief to combat both mechanical problems and environmental effects due to corrosion.

3.5.5 Machining Characteristics

Machinability of an alloy varies with its structure, tendency to work-harden, and response to lubrication and cooling. The severity of various machining operations is rated from mild (e.g., sawing) to severe (e.g., internal broaching, a multiple shaving operation).

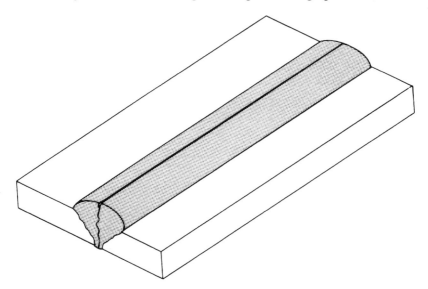

Figure 3.10 Weld cracking.

Sulfur is the element most commonly added to improve the machinability of ferrous alloys. Unfortunately, this leads to hot-short cracking in the free-machining grades of austenitic stainless steels if welding operations are attempted, as in repairing a pump shaft which has cracked from fatigue.

In addition, the iron or manganese sulfides formed can act as focal points for pitting-type corrosion. In a similar fashion, pools of lead in free-machining copper-based alloys are susceptible to localized attack in some environments.

3.5.6   Surface Finish

Finished machined parts in rotating equipment such as pumps and compressors are often polished for improved mechanical operation. The internal surfaces of vessels may be ground or polished to minimize sticking (e.g., in the handling of polymeric resins), although the benefit is often transitory in nature. When surfaces are ground rather than polished, as in finishing of welds in tanks and vessels, high superficial tensile stresses can be induced. Such stresses increase the susceptibility to environmental cracking. Surface grinding should be done with care and, if possible, before the final stress-relieving heat treatment.

3.6   Corrosion Characteristics

The susceptibility of a metal or alloy to the various forms of corrosion, discussed in detail in Chapter 6, obviously enters into the matter of materials selection. It is better that the metal suffer a general or uniform attack at some predictable corrosion rate than undergo localized corrosion such as pitting. A relatively uniform rate of attack permits the intelligent use of a corrosion allowance in the mechanical design of process equipment.

Even more insidious than localized corrosion are subsurface corrosion phenomena (dealloying, intergranular corrosion) and cracking phenomena (corrosion fatigue, environmental cracking).

Various types of phenomena associated with the effects of atomic hydrogen may also be of great concern, depending upon the specific materials and environments.

3.7   Amenability to Corrosion Control

Corrosion control, discussed in more detail in Section V, often affects materials selection. One must be familiar with the measures that may be or are required for a particular material of construction.

Aside from simply accepting some reasonable rate of corrosion (which is the most practical approach in some cases), corrosion control may involve any or all of the five major approaches:

(1)   A change of materials;

(2)   A change in environment;

(3)   Use of barrier coatings;

(4)   Application of electrochemical techniques; and

(5)   Specific design features.

The amenability to corrosion control of the materials selected is an important factor in achieving safe, reliable, and profitable operation.

References

1.   *ASME Boiler and Pressure Vessel Code* (New York, NY: ASME).
2.   Brandt, et al., *Chemical Engineering* 99, 4 (1991).
3.   *Metals & Alloys in the Unified Numbering System* (Warrendale, PA: SAE).

Suggested Resource Information

ASM Handbook, Vol. 6, *Welding, Brazing and Soldering* (Materials Park, OH: ASM International, 1993).

ASM Metals Reference Book, 3rd Ed. (Materials Park, OH: ASM International, 1993).

E. A. Avallone, T. Baumeister III, eds., *Marks' Standard Handbook for Mechanical Engineers*, 9th Ed. (New York, NY: McGraw-Hill, 1987).

Z. D. Jastrzebski, *The Nature and Properties of Engineering Materials*, 2nd Ed. (New York, NY: Wiley-Interscience, 1977).

4

Materials Selection Procedures

The ultimate purpose of this book is to aid the process industry engineer in arriving at the optimum materials of construction for the corrosive service under consideration. Preferably, this should be accomplished with professional help by working with a qualified corrosion engineer.

It must be remembered that the term *corrosive service* holds different connotations for different people. However, whether the problem involves only product contamination or the more aggressive conditions of water-side corrosion or chemical processes, it is a corrosion problem for all concerned. Unfortunately, since many corrosion problems are perceived only after the fact, the engineer or maintenance workers in the plant then become immediately concerned with recognition and understanding of corrosion and metallurgical phenomena, and the economics of repairing, refurbishing, or replacing equipment.

It is only common sense that the major effort toward selection of the best materials of construction and any accompanying corrosion control measures be made in the design stages. This effort properly lies within the province of the professional corrosion/materials engineer, if such is available. However, for companies and plants which do not have such expertise available in-house, it often falls upon the design engineer or even his/her maintenance counterpart, or the production supervisor. Quite often, materials selection decisions for maintenance purposes are made by plant personnel.

4.1　Materials Considerations

Selection of optimum materials of construction involves consideration of a number of materials-related factors, the materials-environmental interactions, and problems inherent in specific types of equipment.

4.1.1   Optimum Materials

The optimum materials are those which provide the lowest cost-to-life ratio, when corrected for tax-rates and accounting procedures (see Chapter 34), and which incorporate certain aspects of safety, health, environmental considerations, and reliability. The desired end result includes profitable operation with a minimum of unscheduled shutdowns, possibly with periodic maintenance or replacement.

4.1.2   Factors in Materials Selection

The major factors in materials selection, such as safety and reliability, cost, fabricability, environmental impact, and energy considerations, have been discussed briefly in the preceding chapters. In addition, one must know and understand the response of the material to corrosion, particularly its susceptibility to localized forms of attack such as pitting, crevice corrosion, or environmental cracking, as described in Section II.

Besides these factors, which are intrinsic to the materials, there are extrinsic factors such as the ASME Boiler and Pressure Vessel Code or other regulatory controls: amenability to one or more of the five types of corrosion control (as required); probable response to fire, explosion, or, temperature or pressure excursions; and economic considerations.

4.2   Materials-Environment Interactions

The characteristics of the basic materials of construction are discussed in detail in Section III, the corrosion characteristics of many common environments are covered in Section IV, and specific corrosion control measures are the subject of Section V.

If the probable behavior of the material under consideration is not known with confidence, it must be ascertained either from the information resources described in Chapter 2 or determined by appropriate laboratory, pilot plant, or field corrosion testing.

4.3   Specific Equipment

Not only do certain material's forms have specific limitations (e.g., difficult to weld or machine, or available only as castings suitable for valves and pumps but not for process vessel fabrication), but also, specific types of process equipment have characteristic corrosion and materials problems inherent in their design.

- Rotating equipment has a greater incidence of problems related to galvanic (i.e., two-metal) corrosion, wear, and fretting because of the necessity of running one metal against another under mechanical loading.

- Heat exchangers have galvanic corrosion problems in many instances, as well as hot-wall or condensation corrosion problems simply by their conditions of operation.

- Scrubbers and distillation or extraction columns have problems of corrosion and materials selection associated with the presence of distinctly different environments at different locations within the same vessel.

- Pumps and valves are not necessarily available in the same materials as associated piping and are subject to a higher incidence and severity of velocity effects.

These and other considerations enter into the final materials selection process.

4.4 Procedures and Communications

Despite the complexity of these factors, which may require a skilled professional to "pull it all together," the biggest problem seems to lie in adequate documentation and communication of the final requirements. New project design, particularly, the specific procedures and their associated documentation can be divided conveniently into three phases: *the definition of technology, the definition of facilities,* and *construction and startup.*

4.4.1 Definition of Technology

The definition of technology (sometimes called "DOT") entails a full description of the process, its chemistry, and its conditions. This is essential for new processes and should be developed as early as possible, in conjunction with the research and development department or other originating sources. (*Note:* While a formal DOT obviously is not required for replication of existing processes, always consider that *exact* replication rarely occurs. Minor changes in operating conditions, or even in the associated utilities, e.g., steam quality or cooling water chemistry, may affect the choice of materials.)

A preliminary materials selection is usually based upon the following considerations:

(1) *Ingredients*—the major and minor constituents of each process stream, including trace contaminants, pH (or total acidity when pH is not meaningful), and degree of aeration;

(2) *Process Changes*—possible deviations from the prevailing chemistry or excursions from the ordinary conditions of temperature or pressure; also, aberrations associated with start-up or shutdown conditions;

(3) *Contaminants*—the possible effects of inadvertent contamination of a feedstock or raw material, or the influence of species entering or concentrating through a recycle stream, condenser leaks, etc;

(4) *Catalysis*—the positive or negative effects of metal ions from corrosion processes that may affect either the chemistry of the process itself or the product quality; and

(5) *Product Quality*—contamination of the final product by even minute amounts of corrosion, thus throwing the product off-specification due to iron contamination, discoloration, haze, etc.

A thorough consideration of these factors will aid in deciding which further corrosion or process investigations might be required before a final decision can be made. Sometimes, the appropriate questions can be resolved through laboratory tests of various kinds. At other times, a *pilot plant* operation might be indicated before one can reasonably start design of a full-scale unit.

4.4.2 Definition of Facilities

The definition of facilities (DOF) establishes the necessary kind, size and number of the various types of equipment and their physical layout at the plant site. A considerable amount of standardization can be achieved in consideration of the plant site, utilities, and auxiliary services. The plant location pre-establishes the atmospheric conditions (which govern the selection of paint systems, insulation, and corrosion-resistant bolting, for example), and the kind and quality of utilities. The latter consists of cooling water, fire-control water, potable water, steam, and steam condensate. Soil conditions establish the corrosion control requirements for underground piping, buried tanks, tank bottoms, etc. By standardizing as much of the materials selection as possible for the specific plant site, the engineer is able to consider in more detail the actual process requirements.

Wastewater streams are a special problem, since they reflect the specific contamination by process effluent streams, as discussed further in Chapter 19. Trace amounts of organic species in wastewater can affect profoundly the behavior of plastics and elastomers used in piping, gaskets, diaphragm valves, etc.

The process requirements demand optimum materials of construction (and related design considerations) for pressure vessels, heat-exchangers, valves, piping, pumps, tanks, and instrumentation. These requirements must be documented adequately in complete flow diagrams called P&ID (process and instrumentation diagrams) or P&CD (process and control diagrams). The P&ID should include reference specifications and standards, as well as process stream compositions and conditions. Also needed are descriptive *equipment lists* and *line lists*, *criteria* (describes what one is trying to accomplish), and *engineering standards* (describes how the criteria are to be effected).

4.4.3 Construction and Startup

The P&ID and related lists and standards must be complete and must adequately transmit the specified requirements to the equipment negotiators, buyers, vendors, fabricators, subcontractors, and inspectors involved in construction and start-up.

To obtain adequate quality assurance, fracture toughness requirements must be established (Chapter 6), inspection methods and criteria spelled out (Chapter 39), and a materials identification procedure (MIP) provided.

The MIP may include alloy verification (by spot-test or other qualitative or quantitative analytical procedures), and such qualification tests as may apply (e.g., the ASTM A 262[1] test for intergranular corrosion in conventional austenitic [type 300 series] stainless steels, the so-called "18-8" stainless steels).

Water quality requirements, temperature limitations, and procedures must be established for hydrostatic testing, "dummy running" (i.e., startup of distillation columns on water only), tank settling, and wet lay-up procedures.

Welding standards may be required, and weld quality and finish requirements are needed, especially for equipment to be coated internally. Over-matching rods (i.e., higher alloy content than the base metal) may be required for alloy construction under certain conditions.

Procedures for pre-service chemical or physical cleaning of equipment must be established, with adequate safety procedures, and provision made for disposal of resultant wastes in accordance with environmental quality requirements.

Assistance should be rendered during start-up as to the proper installation and operation of any necessary corrosion control measures and the monitoring or on-stream inspection (OSI) of the equipment and materials selected.

4.5   Conclusion

The principles enumerated above constitute only a thumbnail sketch of the process of materials selection. More detail can be found in the recommended reading and in the NACE education course *Designing for Corrosion Control–Student Notebook* on this subject.

The following chapters provide insight into the corrosion aspects, materials characteristics, behavior of the major corrosive environments, and the application of specific corrosion control methods.

Reference

1. ASTM Standard A 262, "Standard Practices for Detecting Susceptibility to Intergranular Attack in Austenitic Stainless Steels" (Philadelphia, PA: ASTM).

Suggested Resource Information

C. P. Dillon, "The Role of Communications in Corrosion Control," *Materials Performance* 27, 1 (1988), p. 88.

C. P. Dillon, *Materials Selection for the Chemical Process Industries* (New York, NY: McGraw-Hill, 1992).

R. J. Landrum, *Fundamentals of Designing for Corrosion Control–A Corrosion Aid for the Designer* (Houston, TX: NACE International, 1989).

SECTION II

Corrosion
Considerations

5

Corrosion Mechanisms

There has been as yet no final agreement on the definition of corrosion, at either the national or international level. One school of thought wants to restrict the term to corrosion of metals, in which a transfer of electrons is necessarily involved. A second school wishes to apply the term to any and all materials of construction. The latter school seems to have history in its favor, because the term *corrosion* has been used historically in reference to the deterioration of concrete, for example, as well as of metals and alloys. For our purposes, I have elected to stay with the second school of thought.

5.1   Definition

Corrosion is the deterioration of a material of construction *or of its properties* as the result of exposure to an environment. The term *deterioration of properties* is used in the definition because there may be, for example, a loss of ductility without any material loss or dimensional changes. Corrosion may entail loss of mass, gain in weight, or changes in physical or mechanical properties. Sometimes, there may be no obvious change(s) in the appearance of the artifact.

5.2   Electrochemistry of Metallic Corrosion

5.2.1   Fundamentals

Corrosion entails the conversion of a metal from the atomic to the ionic state, with the loss of one or more electrons. By definition, this *anodic* reaction produces a positively charged metal ion and free electrons in accordance with the following reaction:

$$M^\circ \rightarrow M^{n+} + ne \qquad (1)$$

for example,

$$Fe° \rightarrow Fe^{2+} + 2e \qquad (2)$$

From the chemical standpoint, corrosion is, by definition, primarily the anodic reaction (i.e., the loss of metal) and an *oxidizing process* because it involves the loss of electrons. However, changes in properties due to corrosion (e.g., hydrogen embrittlement [HE]), as well as some specific corrosion phenomena (e.g., sulfide stress cracking [SSC]), result from the concurrent cathodic reaction.

Just as a battery will not function until the two terminals (positive and negative) are connected through an external circuit, the anodic reaction cannot proceed without a corresponding *cathodic* reaction, which is a *reduction* process. The electrons released by the anode travel through the external circuit and react with some species at the surface of the cathode. In aqueous solutions, this reaction is (3) the reduction of hydrogen ions (i.e., protons) to atomic hydrogen, or (4) reduction of both water and dissolved oxygen to hydroxyl ions:

$$H^+ + ½ O_2 + e \rightarrow H° \qquad (3)$$

$$H_2O + ½ O_2 + 2e \rightarrow 2OH^- \qquad (4)$$

When the aqueous solution is acidic, protons reduce dissolved oxygen to water:

$$2 H^+ + ½ O_2 + 2e \rightarrow H_2O \qquad (5)$$

The driving force that makes metals corrode arises from the large energy input required to smelt them from their ores to the metallic state. The order of corrosion resistance to natural environments can be defined roughly by the relative tendency to exist as a metal in the natural state. Gold and silver are found free in various geographical locations, but iron is found only as the ore (e.g., iron oxide, also known as *hematite*, Fe_2O_3).

Table 5.1 lists some commonly used metals in order of diminishing amounts of energy per mol required to reduce them from their ores. The more reactive or anodic materials are at the top, while the less reactive ("noble") metals are at the bottom of the list.

Table 5.1

Metals by Order of Energy/Mol
Required for Smelting

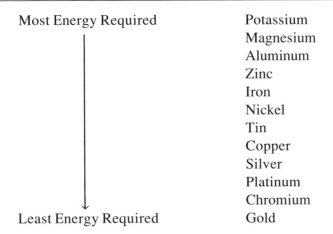

Most Energy Required	Potassium
	Magnesium
	Aluminum
	Zinc
	Iron
	Nickel
	Tin
	Copper
	Silver
	Platinum
	Chromium
Least Energy Required	Gold

As with the dry cell or battery, there are five conditions which *must be met* before corrosion can occur:

(1) An electrical potential difference (i.e., voltage) between the anode and cathode to drive the reaction;

(2) An anodic reaction;

(3) An equivalent cathodic reaction;

(4) An electrolyte present for the internal circuit (i.e., an environment which will conduct electricity, such as saline water); and

(5) An external connection or circuit (i.e., a direct electrical contact) between the anode and cathode. (In single-metal corrosion, the metal itself constitutes the external circuit between discrete anodes and cathodes, as described further below.)

Even when all these conditions are fulfilled, corrosion may be stifled by polarization.

Polarization is a change in potential as the result of current flow (Figure 5.1). Either the anodic or cathodic reaction (or both) may become polarized, but cathodic polarization is more common in natural environments, like soil or water. Anodic polarization occurs when the corrosion products are insoluble in the environment (e.g., an accumulation of lead sulfate in lead immersed in dilute sulfuric acid). An example of cathodic polarization is the accretion of molecular hydrogen at the cathode (as with steel in deaerated water); the hydrogen neither being evolved (as in dilute acid) nor consumed by reduction of dissolved oxygen (as in aerated water). The accumulation of unreacted hydrogen as a film not only physically retards current flow, changing the internal circuit resistance, but also shifts the potential of the cathode.

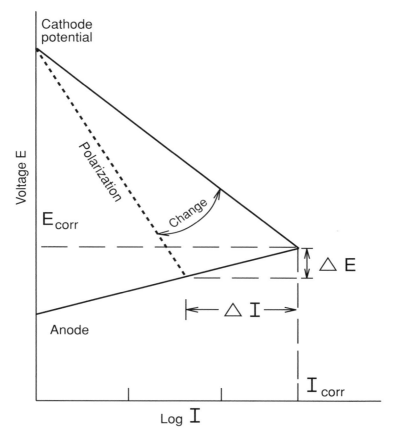

Figure 5.1 Polarization diagram.

5.2.2   Single-Metal Corrosion

When a single metal or alloy corrodes in a homogeneous environment, the discrete anodes and cathodes are microscopically tiny spots on the metal surface. Any particular site does not remain an anode or cathode, but changes back and forth from one capacity to another, providing an overall effect of a uniformly corroding surface. Under these circumstances, the factors and functions analogous to an electrical battery are as follows:

> *Voltage*—the difference in energy levels between discrete anode and cathode sites on the metal surface;
>
> *Electrodes*—specific discrete anode and cathode sites;
>
> *Electrolyte*—the corrosive environment; and
>
> *External Circuit*—the metallic continuity between the anode and cathode sites.

Even with a single metal, there may be abnormal potential differences between specific sites. These can arise from differences in metallurgical structure, from different degrees of cold-work, or from different levels of residual stress. In such cases, there exists a *nonuniform* metal surface exposed to the environment, with the potential difference driving the corrosion reaction.

In other cases, part of the metal surface may either have or acquire a surface film with a different solution potential than the metal itself. (As we will see below, this condition becomes similar to two-metal, otherwise known as bimetallic or *galvanic* corrosion.) This film may be the result of prior treatment, such as the mill-scale or "magnetite" film, Fe_3O_4, on hot-rolled steel, or can develop in service, such as the lead sulfate film on lead or the complex oxides on copper alloys.

It also is possible to have a uniform metal surface but a non-homogeneous environment. A common example is the oxygen concentration cell, where oxygen becomes depleted in a crevice or other shielded area (e.g., under deposits). Such a cell can be demonstrated easily in the laboratory by immersing twin steel electrodes in an aqueous solution divided by a semipermeable membrane (Figure 5.2). When air is bubbled through one segment and nitrogen through the other, a voltage difference, or current

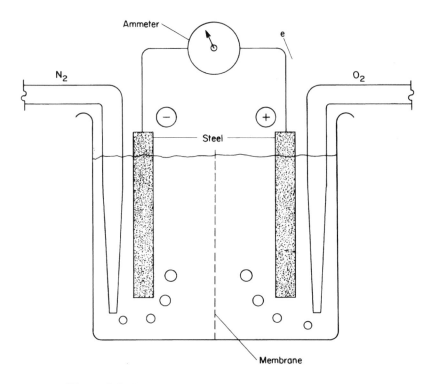

Figure 5.2 Oxygen concentration cell; lab demonstration.

flow, can be measured, and it will be noted that the oxygen-starved electrode is the anode. Note that corrosion occurs where the oxygen is absent. Although essential to the corrosion mechanism, the role of oxygen is to function as a cathodic depolarizer; the oxygen-rich electrode is the cathode and does *not* corrode.

5.2.3 Two-Metal (Galvanic) Corrosion

The accelerated attack which occurs on one metal as a consequence of electrical contact with another in a corrosive environment is a matter of common observation. More than 200 years ago, Sir Humphrey Davy reported accelerated corrosion of rudder irons as a consequence of installing copper sheathing on the wooden hulls to protect them from marine borers. Aluminum screens suffer corrosion if installed with steel or brass screws. Cast iron water-boxes on copper alloy condensers suffer accelerated attack in cooling water.

When extraneous factors are held constant, the difference in corrosion tendency between different metals can be determined quantitatively by measuring their individual potentials against a reference half-cell electrode. (A standard half-cell is one with reproducible open-circuit potential, e.g., a calomel or mercury-mercuric chloride electrode, a silver-silver chloride electrode, a copper-copper sulfate electrode, etc.) (See Figure 5.3.) The potentials so measured are often corrected to a value against a theoretical standard (or normal) hydrogen electrode (SHE).

When the open-circuit potential is determined under standard conditions, i.e., for a pure metal in a 1 molar (M) solution of its own ions at 25°C (77°F), the electromotive series is obtained (Table 5.2). These reversible reactions between a pure metal and its own ions allow one to rate metals in order of decreasing (less negative) activity.

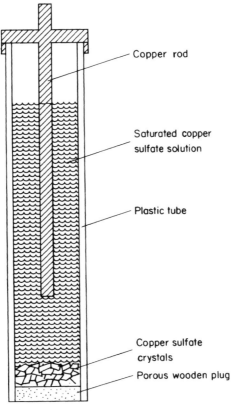

Figure 5.3   Copper–copper sulfate reference half-cell.

Table 5.2
Electromotive Series[1]

Metal-metal ion Equilibrium (Unit Activity)		Electrode Potential vs Normal Hydrogen Electrode at 25°C, volts
Active	$K\text{-}K^+$	-2.925
(anodic)	$Na\text{-}Na^+$	-2.714
	$Mg\text{-}Mg^{2+}$	-2.363
	$Al\text{-}Al^{3+}$	-1.662
	$Ti\text{-}Ti^{2+}$	-1.630
	$Zn\text{-}Zn^{2+}$	-0.763
	$Cr\text{-}Cr^{3+}$	-0.744
	$Fe\text{-}Fe^{2+}$	-0.440
	$Cd\text{-}Cd^{2+}$	-0.403
	$Co\text{-}Co^{2+}$	-0.277
	$Ni\text{-}Ni^{2+}$	-0.250
	$Sn\text{-}Sn^{2+}$	-0.136
	$Pb\text{-}Pb^{2+}$	-0.126
	$H_2\text{-}H^+$	$+0.000$
	$Cu\text{-}Cu^{2+}$	$+0.337$
	$Hg\text{-}Hg^{2+}$	$+0.788$
Noble	$Ag\text{-}Ag^+$	$+0.799$
(cathodic)	$Pd\text{-}Pd^{2+}$	$+0.987$
	$Pt\text{-}Pt^{2+}$	$+1.200$
	$Au\text{-}Au^{3+}$	$+1.498$

[1] M. G. Fontana, *Corrosion Engineering*, 3rd Ed. (New York, NY: McGraw-Hill, 1986), p. 42.

This series predicts the tendency of one metal in ionic form to displace another in aqueous environments. The fact that one metal is more noble or less electronegative (i.e., cathodic) than another is easily demonstrated. For example, a piece of steel dipped in a copper sulfate solution acquires a flash copper coating. The steel suffers corrosion, but an electrochemical equivalent of copper is plated out by a process known as *cementation*. Likewise, copper will plate out on aluminum in aqueous solutions and silver on copper from a silver nitrate solution. In each case, the more anodic material suffers corrosion and releases its ions, with the ions of the more noble metal (cathodic) metal plating out in the process. The anodic reaction is the oxidation of the base metal; the cathodic reaction is the reduction of ions of the more noble metal.

When metals and alloys are arranged in order of diminishing negative potential in a real-world situation such as in seawater, a galvanic series is obtained (Table 5.3). Such a series can be constructed by immersing different metals in a liquid environment and rating them in order of increasing corrosion resistance. Such a series also can be constructed in a more quantitative manner by measuring the open-circuit potential, using a high-resistance volt-meter, against a standard reference half-cell electrode. Those readings can be converted to the values against a standard hydro-gen electrode (SHE), as demonstrated in Table 5.3. The driving voltage of a galvanic couple is the difference between the open-circuit potentials of the members. For example, there is approxi-mately 0.4 volts potential between steel and copper in seawater.

Table 5.3

Galvanic Series in Seawater

	Metal or Alloy	Voltage vs SHE[(A)]
Active (anodic)	Magnesium	− 1.49
	Zinc	− 0.81
	Cadmium	− 0.64
	Aluminum	− 0.61
	Steel	− 0.38
	Type 300 Series SS (active)	− 0.36
	Lead	− 0.32
	Tin	− 0.27
	Admiralty Metal	− 0.12
	Hydrogen	0.00
	Copper	+ 0.02
	Nickel	+ 0.10
	Monel	+ 0.13
	Titanium	+ 0.14
	Type 300 Series SS (passive)	+ 0.15
	Silver	+ 0.16
	Graphite	− 0.49
	Platinum	+ 0.50
Noble (cathodic)	Gold	+ 0.50

[(A)] To convert SHE to saturated calomel electrode (SCE), add -0.24.

The galvanic series, in common practice, is established for seawater. However, it should be noted, that while this galvanic series is generally applicable in many natural waters, it does not necessarily apply. In waters of certain chemistry, for example, zinc and steel can reverse their relative positions at moderately elevated temperatures. Also, significantly different chemical environments will have their own specific galvanic series.

It is interesting to note the different positions of titanium, for example, in the electromotive force (emf) vs the galvanic series in seawater. Titanium is strongly anodic in the emf series because of its reactive nature. It is one of the strongest cathodes in the galvanic series because of its rapid polarization by formation of a *passive* film. From a practical standpoint, its behavior in the real world of aqueous solutions is more relevant.

Passivity is a condition in which a relatively active metal assumes more cathodic, or noble, characteristics, either temporarily (as with steel in strong nitric or sulfuric acid) or permanently by virtue of a film of oxidized metal or alloy (e.g., aluminum, titanium, stainless steels). The word *permanently* is somewhat misleading because the protective oxide film can be removed either chemically or mechanically, leaving the metal in an active (corrodible) condition. This is why the stainless steels are assigned two positions in the galvanic series—one as active and one as passive—although the latter is the more prevailing condition, except under most unusual circumstances. Indeed, the stainless steels are self-passivating in aerated waters.

The fundamental relationships in a galvanic cell are expressed by Ohm's Law:

$$E = IR \qquad (6)$$

where E is the voltage in volts, I is the current in amperes, and R is the resistance in ohms.

If the current flowing in a galvanic cell is measured, it will be proportional to the amount of metal corroded over unit time, in accordance with Faraday's Law:

$$W = ITZ \qquad (7)$$

where W is the loss of mass, I is the current in amperes, T is the time, and Z is the electrochemical equivalent of the metal, proportional to strength of current flow; duration of current flow; and the chemical equivalent weight of the corroding metal.

In any galvanic couple, the metal near the top of the series will be the anode, suffering accelerated attack, while the other will be the cathode and be protected, suffering less corrosion than it would as an individual, freely-corroding metal. This beneficial effect is called *cathodic protection* (CP), and is discussed briefly in Chapter 5.2.4 and in detail in Chapter 38.

Assuming bimetallic contact in a corrosive environment, three major factors control the actual amount of corrosion: the relative areas of the anode and cathode—the geometry of the physical layout and internal circuit—and polarization. (*Note*: The resistance of the electrolyte is also an important factor, as discussed in the consideration of corrosion in soil in Chapter 20. However, it is of less practical importance under immersion conditions in electrolytes.)

In accordance with Faraday's law, the amount of current determines the amount of anodic metal which will be dissolved. (It happens that 1 amp of DC flowing for 1 year will dissolve approximately 20 lbs [9 kg] of steel, as will 8,760 amps flowing for 1 hour.) In most cases of galvanic corrosion in natural environments, the system is under *cathodic control*, i.e., the amount of current is determined by the area of the cathode. Consequently, even under total immersion conditions, it is not particularly harmful to put bronze bolts in a large steel plate, whereas a steel rivet in a copper plate will be corroded very rapidly. (This effect would be less noticeable in low-conductivity steam condensate than in highly conductive seawater, as mentioned above.) It is axiomatic that high cathode-to-anode ratios should be avoided wherever possible.

The geometry of a galvanic couple can affect current flow because the current will flow through the easiest path. For example, the galvanic corrosion of a steel or cast iron waterbox (i.e., bonnet or head) is caused by the bronze tubesheet and the tube ends. Only the first four to six tube diameters of copper alloy tubes can enter into the circuitry of the couple. The rest of the tube internal diameter is too far away, electrically speaking, to enter into the circuitry (Figure 5.4). (*Note*: This does not hold true for

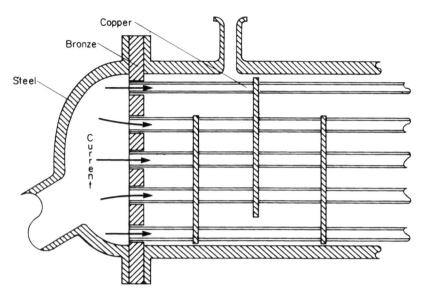

Figure 5.4 Geometric effects in galvanic corrosion of steel waterbox (bonnet or head) by copper alloy tubes and sheet.

materials like high-performance stainless and related alloys or for titanium, where the *entire* tube-length constitutes an effective cathode.)

Normally, the controlling cathode is depolarized (e.g., by dissolved oxygen), allowing corrosion to proceed as governed by the factors described. However, in the absence of cathodic depolarizers, hydrogen will accumulate on the cathode, setting up increased resistance as well as a back emf, and diminishing the galvanic corrosion. No significant corrosion will occur at ambient temperatures in anaerobic seawater, for example, barring other depolarization effects. Cathodes also are polarized frequently in water by calcareous deposits (i.e., calcium and magnesium salts), as discussed in Chapter 19 on water chemistry.

It should be emphasized once again that each chemical environment (if significantly different from water or from dilute aqueous solutions) will have *its own galvanic series.* The relationships shown in the galvanic series in seawater (Table 5.3, page 63) may become quite different in other environments. Magnesium is cathodic to steel in hydrofluoric acid; steel to stainless steel in hot caustic; and, zinc to steel in hot water of specific chemistries.

5.2.4   Cathodic Protection (CP)

Cathodic protection (CP) is a practical utilization of galvanic corrosion. In the early 1940s, an investigation was made as to the practicality of protecting steel against corrosion in seawater by attaching zinc or magnesium anodes to it. The inexpensive sacrificial anodes would corrode preferentially, thus affording CP to the steel. In so doing, the current generated also precipitated calcareous deposits from the seawater, building a partially protective coating and effectively reducing the area of the steel cathode, thereby limiting the current demand and prolonging the life of the anodes. The investigation also showed that the same effect could be achieved by supplying DC from an external source such as a battery, rectifier, or generator, using either scrap metal or long-lived conductive anodes of carbon or silicon cast iron, for example.

The major impetus for CP came after World War II, when the process was commercially applied to underground pipelines. Underground steel tends to be corroded by low-resistivity (i.e., high conductivity) soils, as described in Chapter 20.

From a practical standpoint, the steel should be coated or wrapped to reduce the cathode area (and current demand) to that required by "holidays" (i.e., breaks or defects) in the coating. The steel then can be protected economically either by sacrificial anodes or by driven anodes using an external DC source. In practice, DC is produced by rectifiers which convert AC to DC. The choice between sacrificial and driven anodes systems will be dictated by economics, geometry, and possible damage to surrounding structures by stray currents or interference (i.e., so-called electrolysis).

Electrolysis is corrosion of a buried or immersed structure due to stray DC from some external source, such as an electric motor or welding machine. The current leakage from a DC motor, welding machine or rectifier will corrode steel at approximately 20 lbs (9 kg) per amp-year. Naturally, the smaller the anodic area from which the steel is lost, the higher the corrosion rate at that location (e.g., holidays [defects] in well-coated lines).

The problem was first observed around electrical railways where DC was supplied from an overhead wire (or "trolley") and was supposed to return to its source along the track or third rail. If the current found it easier to return to its source on buried steel pipe (e.g., a water line or sewer system), it would protect the pipe at the point of entry but cause severe corrosion at the point of discharge via the soil to the power source. Similarly, ship hulls have been damaged during reverse-polarity welding operations, when the welding machine was ashore and inadequate electrical returns were provided, or where the return cable connections were faulty. In either case, the current simply found it easier to return to its source through the hull-seawater interface than through the return leads, causing severe hull damage. Welding austenitic stainless steel pipe in place for underground installations has resulted in similar problems in the form of severe localized pitting.

5.2.5   Anodic Protection (AP)

There are particular cases in which current flow *from* the metal can be beneficial, rather than causing electrolysis. Some particular combinations of metals and environments lend themselves to *anodic protection* (AP). The situation requires that, for its corrosion resistance, the anodic metal be dependent upon an insoluble film which can be reinforced and maintained by the anodic effect of an impressed anodic polarization.

The classical examples of AP are in storage of steel storage tanks for concentrated sulfuric acid (to reduce iron contamination of the acid) and in stainless steel coolers for strong acid. AP *absolutely* requires professional design and operating supervision, as corrosion by electrolysis may ensue if the proper potentials are not maintained continuously. (Obviously, power failures will permit corrosion to proceed when external sources are employed for either CP or AP.)

The advent of microcomputers has expanded the application of electrochemical corrosion control. Not only are CP and AP handled in a more sophisticated manner through computer control, but also, other applications are also in use, e.g., the prevention of chloride effects (pitting or stress corrosion cracking in pulp and paper plants by constant control of the open-circuit potential of the austenitic stainless steel equipment).

Suggested Resource Information

M.G. Fontana, *Corrosion Engineering,* 3rd Ed. (New York, NY: McGraw-Hill, 1986).

6

Corrosion and Metallurgical Phenomena

The ability to recognize and distinguish between different corrosion, metallurgical, and mechanical phenomena is an important part of failure analysis and, by extension, of corrosion control decisions.

In this chapter, we will discuss the three groups of phenomena in an introductory manner. Certain sub-groups are of sufficient importance to warrant more detailed discussion in their own right, and are covered in more detail in subsequent chapters. Inevitably, as noted in the discussion below, there is a certain amount of overlap between some types of phenomena.

6.1   Corrosion Phenomena

Although there is not total agreement even among experts, it is generally accepted that there are basically eight forms of corrosion. They are derived from those originally expounded by M.G. Fontana, and are as follows: [1, 2]

(1)   General (uniform) corrosion;

(2)   Localized corrosion;

(3)   Galvanic corrosion;

(4)   Cracking phenomena;

(5)   Velocity effects (erosion-corrosion, cavitation, fretting);

(6)   Intergranular attack (IGA);

(7)   Dealloying (parting corrosion); and

(8)   High-temperature corrosion.

These forms of corrosion can be divided into three categories:

Group I — Those readily identified upon visual examination (Forms 1, 2, and 3);

Group II — Those which may require supplementary means of examination (Forms 5, 6, and 7); and

Group III — Those which usually should be verified by microscopy, optical, or SEM, although sometimes apparent to the naked eye (Forms 4 and 8).

The individual phenomena are illustrated schematically in Figure 6.1 and defined further in Chapter 6.1.1.

6.1.1 Definitions

Group I Forms

(1) *General Corrosion.* General corrosion is characterized by an even, regular loss of metal from the corroding surface. All metals are subject to this type of corrosion under some conditions, such as atmospheric rusting of or dissolution of zinc by dilute acid. It is the most desirable form of attack because it lends itself to predicting the life of equipment. It is observed in both single-metal and bimetallic corrosion (Form 3 below).

(2) *Localized Corrosion.* Localized corrosion is that in which all or most of the attack occurs at discrete areas. These may be relatively large but shallow, as with the "corrosion lakes" encountered in oil and gas equipment. However, *pitting*, with its intense localized attack is of greater consequence. Characteristically, pitting is at least as deep as it is wide.

Crevice corrosion is a particular form of pitting which occurs between adjoining surfaces (e.g., in threaded connections or flanged connections), usually due to oxygen concentration cell effects.

(3) *Galvanic Corrosion.* Also known as bimetallic or two-metal corrosion, galvanic corrosion is accelerated attack occasioned by electrical contact between dissimilar conductors in an electrolyte. The accelerated corrosion is suffered by the anodic member of the couple (e.g., steel in the iron-copper cell), and

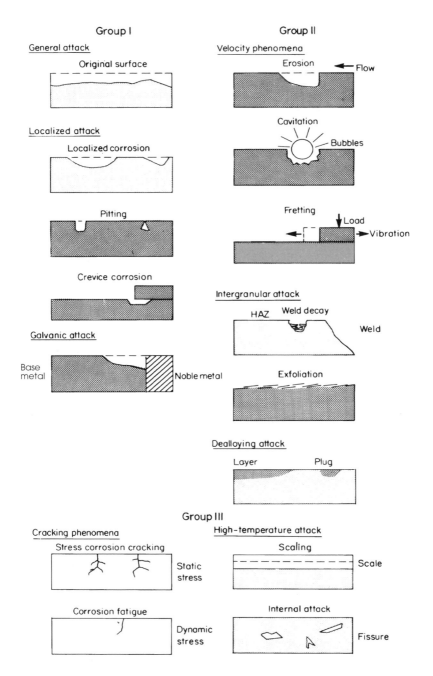

Figure 6.1 Eight forms of corrosion (mechanics of attack). (From NACE International Handbook 1, *The Forms of Corrosion Recognition and Prevention. Note:* Base metal corrosion of galvanic attack has been modified. Reproduced with permission.)

depends upon their relative position in the pertinent galvanic series and the relative areas and geometry of the two electrodes. Galvanic corrosion also can be caused by conductive films such as mill-scale and by conductive nonmetallic cathodes such as carbon or graphite.

Group II Forms

(4) *Velocity Effects*. Velocity effects comprise erosion-corrosion, cavitation and fretting. *Erosion-corrosion* is attack accelerated by high-velocity flow or impingement. There is actually a family of phenomena ranging from purely mechanical effects such as abrasion and wear to the corrosion-related effect in which an otherwise protective film of corrosion products is swept away by the flow conditions. In all cases, there is a distinctive flow pattern (e.g., copper in water, steel in steam).

Cavitation is a distinctive form of velocity attack caused by the implosion of bubbles formed where the local pressure in the flowing liquid drops below the vapor pressure. One observes a porous, gouged effect quite unlike the flow-lines of erosion-corrosion. Cavitation is observed on ship propellers, pump impellers and on the internal bore of water-cooled compressor rods. Its mechanical aspects can be demonstrated (e.g., by ultrasonic vibration of glass in distilled water), as can its electrochemical features (e.g., response to corrosion inhibition and CP).

Fretting is another form of corrosion associated with motion, in this case between mating surfaces under load and subjected to vibration. This induces the tearing away of small particles which are subsequently oxidized or otherwise corroded between the facing surfaces. A typical example is the fretting of automotive wheel bearings if the vehicles are inadequately supported during rail shipment.

(5) *Intergranular Attack*. IGA is preferential attack at the grain boundaries of a metallic structure. In some cases, whole grains may fall out, e.g., "sugaring." The attack may be general if an artifact is improperly heat-treated, or localized in heat-affected zones (HAZ) of welds, for example, the so-called "weld decay" of certain austenitic stainless steels.

(6) *Dealloying Corrosion*. Dealloying or "parting" corrosion is the selective removal of one metallic constituent of an alloy. It is exemplified by the dezincification of yellow brass. (*Note:*

Layer-type dezincification corresponds to uniform corrosion while *plug-type* corresponds to pitting.) Another manifestation is the graphitic corrosion (sometimes erroneously called *graphitization*) of gray cast iron. Typically, the selective dissolution of an alloy constituent seriously weakens the metal structure without necessarily changing the apparent physical dimensions.

Group III Forms

(7) *Cracking Phenomena.* The cracking phenomena include both the corrosion-related environmental cracking and the mechanical-electrochemical fatigue. *Environmental cracking* comprises three forms of attack which involve the brittle cracking of a metal or alloy under the combined effect of a tensile stress and a *specific* corrodent. The three forms are:

(a) *Stress Corrosion Cracking* (SCC). An anodic process exemplified by the chloride SCC of stainless steel. (*Note:* As discussed further in Chapter 18, the term *Environmental Stress Cracking (ESC)* is used for the cracking of plastic materials.);

(b) *Hydrogen-Assisted Cracking* (HAC). A cathodic process exemplified by sulfide stress cracking (SSC) of hardened steel by hydrogen sulfide; and

(c) *Liquid Metal Cracking* (LMC). Also called liquid metal embrittlement (LME). This is a fissuring process exemplified by the action of mercury on cold-worked brass or molten zinc on austenitic stainless steels.

Fatigue is the brittle cracking of an otherwise ductile material induced by numerous repetitions of a cyclic stress. It becomes *corrosion fatigue* when a *nonspecific* corrodent aggravates the situation and eliminates the endurance limit, as previously described in Chapter 3.

(8) *High Temperature Corrosion.* This includes a variety of phenomena which involve conversion of the metal to a metallic compound. The most common product is the metal oxide, but other conversions (e.g., halides, sulfides, carbides, nitrides, etc.) are possible. Reactions may occur on the metal surface or within the structure of the metal itself, as discussed further in Chapter 31.

6.2   Metallurgical Phenomena

There are two metallurgical phenomena which, although not corrosion-related, are nevertheless important in materials selection.

6.2.1   Nil Ductility Effects

Nil ductility is the phenomenon in which an otherwise ductile metal or alloy becomes distinctly brittle with decreasing temperature, as discussed briefly in Chapter 3.2.4, and shown in Figure 3.2. Steels are particularly susceptible to this phenomenon, but so are such corrosion-resistant alloys as the low-interstitial ferritic stainless steels. The exact nil ductility transition temperature (NDTT) will vary primarily with thickness of the material but also is influenced by compositional variables and heat treatment. Special impact tests have been devised which are used to determine the NDTT for a particular set of conditions, including flaw size (which can be very critical). This is an area in which advice from specialists is required.

Note that this is a problem distinct from the behavior of such inherently brittle alloys as cast iron or 14.5% silicon iron (UNS F47003).

6.2.2   Hot-Short Cracking

Hot-short cracking is the microfissuring which occurs under tensile stress at elevated temperatures, as when welding under restraint, due to low-melting constituents at the grain boundaries. When a metal is hot-worked or welded, these low-melting constituents separate (even though the metal itself is well below its melting range) and leave a network of microfissures. These usually develop quickly into visually detectable macrocracks, as well. Hot-short cracking bears a distinct resemblance, upon microscopic examination, to environmental cracking in an intergranular mode. However, it is usually detectable immediately after occurrence by such methods as dye penetrant or ultrasonic inspection. Columbium-stabilized stainless steels, silicon-bronzes, and nickel-molybdenum alloys are particularly susceptible to hot-short cracking.

A more detailed discussion of the most important phenomena will be found in subsequent chapters. Although passing reference has been made to specific materials of construction, detailed descriptions of materials are provided in Section III.

References

1. M. G. Fontana, ed., "The Eight Forms of Corrosion," *Process Industries Corrosion* (Houston, TX: NACE International, 1975), p. 1.

2. C. P. Dillon, *NACE Handbook 1–Forms of Corrosion: Recognition and Prevention* (Houston, TX: NACE International, 1982), p. 3.

Suggested Resource Information

M. G. Fontana, *Corrosion Engineering*, 3rd Ed. (New York, NY: McGraw-Hill, 1986).

7

Sensitization and Weld Decay

Two topics of special interest are sensitization and weld decay, because of the unhappy consequences of intergranular corrosion in process equipment. Together with environmental cracking (Chapter 8), they are the primary concern in many corrosion applications.

The term *sensitization* is used broadly to describe the susceptibility to IGA which results from exposing a metal or alloy to a critical temperature range (as in heat treatment or welding). Sensitization may be general (as when an entire item is improperly heat-treated) or localized (as by spot-heating or welding).

The most common form of localized sensitization occurs in the HAZ adjacent to welds which results from the temperature gradient between the cold parent metal and the molten weld. The resultant IGA of the HAZ in certain environments is commonly known by the misnomer, "weld decay." Such a term should have been reserved for those incidents in which the weld *itself* is selectively corroded, as with type 316L in some organic acid services. Figure 7.1 shows a schematic diagram of the location and nature of the so-called weld decay in a HAZ.

The problems of sensitization and weld decay are experienced most commonly in the austenitic stainless steels (see Chapter 12), as discussed further below. However, they are also encountered in certain high-nickel alloys, and occasionally in other metallic materials.

When an austenitic stainless steel of the type 300 series with moderate carbon content (e.g., type 304; 0.08% carbon maximum [UNS S30400]), is heated in the temperature range from 425°C (800°F) to 815°C (1,500°F), the dissolved carbon migrates to the grain boundaries and precipitates as chromium carbides. This leaves a chromium-depleted zone of diminished corrosion resistance around each individual grain, resulting in a susceptibility to IGA *in certain environments* (not all, by any means). Exposure to this critical sensitizing temperature range can be the result of

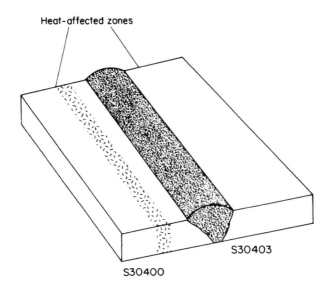

Figure 7.1 Schematic of weld decay (IGC) of 18-8 stainless steel.

improper annealing, inadequate quenching from a solution-anneal, slow cooling during thermal stress-relief, hot-working (e.g., spinning a dished head), or from heating in welding or forming operations.

It is important to note that the environment must be specific for this form of attack. Oxidizing acids like nitric acid and acids containing oxidizing agents (e.g., sulfuric or phosphoric acid containing ferric or cupric ions), are highly specific for this type of attack. So too are hot organic acids such as acetic acid and formic acid. Seawater and other high chloride waters cause severe pitting in sensitized areas, but low-chloride waters (e.g., potable water) do not, except in particular situations, for example, as might occur under the influence of bacterial action.

Originally, sensitization and weld decay of the type 300 series stainless steels were combatted by alloy additions of titanium, columbium (niobium), and columbium-tantalum mixtures. This practice continues today to some extent and is also employed with higher alloys like the high-performance stainless steels and nickel-rich alloys (although lowering the carbon content is also utilized, as discussed further below).

Adding titanium or columbium in amounts equal to five or ten times the carbon content, respectively, permits the alloy to precipitate titanium or columbium carbides (in lieu of chromium carbides) during sensitizing heat exposures. The carbon is still precipitated, but without diminishing the chromium content at the grain boundaries. Types 321 (Ti-stabilized [S32100]) and 347 (Cb-stabilized [S34700]) were thus derived from type 304.

Two problems exist with this approach. Titanium does not transfer well across a welding arc, thus losing much of its effectiveness in multi-pass welding or cross-welding (intercepting vertical and horizontal welds). Although columbium does not have this deficiency, the columbium carbides (as well as titanium carbides) can be redissolved by the heat of welding, especially with alloys of higher nickel content. Consequently, multiple-pass welding or cross-welding can first redissolve titanium or columbium carbides, and then permit chromium carbide precipitation in the *fusion zone* (not the HAZ). This can result in a highly localized form of intergranular corrosion known as *knife-line attack* (KLA), observed particularly in alloys like type 347 and alloy 825 (Ni-Fe-Cr alloy, UNS N08825).

With the development first of low-carbon ferrochromium and later with modern improved steel-making processes, it became possible to manufacture type 300 series stainless steels of much lower carbon content. When the carbon content is held to less than approximately 0.030%, chromium carbide precipitation can still occur upon sensitization, but in such small amounts that no significant chromium depletion occurs. Modern low-carbon alloys (e.g., types 304L, 316L, and 317L) practically are immune to weld decay. They will, however, become sensitized upon *prolonged* heating in the sensitizing range, such as in service, or during very prolonged thermal stress relief. Graphs are available which depict the time-temperature-transition characteristics of such alloys. From a practical standpoint, they can usually be welded, hot-formed, and even thermally stress-relieved without sensitization occurring.

The newer, high-performance stainless steels and nickel-rich alloys often combine stabilization (e.g., titanium or columbium additions) and very low carbon contents. The high-nickel alloys (e.g., Ni-Cr-Mo alloy C-276 [N10276]) combine carbon control with specific ratios of other critical elements.

Both composition and thermal history may require verification other than that provided by the manufacturers' records. There are specific quality assurance tests developed and standardized for the evaluation of susceptibility to IGA. Note that these are more reliable than chemical analyses (which only show composition), because they verify the proper *distribution* of the alloying elements and their derivative species. For the type 300 series stainless steels, these tests are exemplified by ASTM A 262[1], with its various practices. Practice A is a screening test, employing an electrolytic oxalic acid etch combined with metallographic examination. The other practices involve exposure, after sensitization, to boiling 65% nitric acid, 10% nitric-3% hydrofluoric acid, acidified ferric sulfate or acidified copper sulfate plus copper, depending upon specific alloys and applications of interest. Similar ASTM tests have been developed for other, higher-alloyed stainless steels, ferritic stainless steels, high-nickel alloys, etc.

Although, in common parlance, the terms *sensitization* and *weld decay* have been used primarily in regard to stainless steels, it should be noted that other metals and alloys are susceptible to IGA as a result of either carbide or other intermetallic compound precipitation, or to pre-precipitation effects.

Aluminum alloys may suffer a type of intergranular corrosion known as *exfoliation*, because of the leaf-like appearance of the surface when corrosion proceeds along the grain boundaries in the direction of cold-working. Aluminum alloys may actually swell, increasing in thickness because of the retention of high-volume corrosion products within the metallic structure.

The weld HAZ of carbon steel piping can be attacked in strong sulfuric acid. Admiralty brass (C44300) may suffer IGA in certain types of water. Monel[†] (Ni-30% Cu, alloy 400) may suffer IGA in hydrofluoric or chromic acids. Alloys B (Ni-28% Mo, N10001) and C (Ni-Cr-Mo, N10002), which are now both obsolete and replaced by the variant alloys B-2 (N10665) and C-276 (N10276) resistant to this phenomenon, suffered IGA due to the precipitation of molybdenum-rich phases. Alloys 600 (N06600) and 625 (N06625) are susceptible in some environments. Zirconium (R60702) suffers weld decay in hydrochloric acid or sulfuric acid environments contaminated with ferric ions.

[†] Trade name

Reference

1. ASTM Standard A 262, "Standard Practices for Detecting Susceptibility to Intergranular Attack in Austenitic Stainless Steels" (Philadelphia, PA: ASTM).

8

Environmental Cracking

Environmental cracking is a form of localized corrosion in which tensile stress and corrosion have a mutually accelerating effect, causing cracking of a susceptible material in a specific environment. The failure entails *brittle* cracking, usually of an otherwise ductile material. Environmental cracking is *not* confined to metals and alloys. Many plastics can suffer this type of attack.

8.1   Metallic Materials

Almost all metals and alloys are subject to environmental cracking in certain environments. The *specificity* between material and environment is a key feature, unlike corrosion fatigue (which lacks this relationship). Also, unlike fatigue, the stresses are most often of a static rather than a dynamic nature. (However, cyclic stresses can cause environmental cracking, or environmental cracking can initiate fatigue in the presence of cyclic stresses.)

Environmental cracking of metals and alloys can be divided into three categories:

(1) *Stress Corrosion Cracking* (SCC). This is an anodic process (i.e., one which can be alleviated by cathodic protection [CP]), exemplified by the chloride SCC of austenitic stainless steels;

(2) *Hydrogen-Assisted Cracking* (HAC). This is a cathodic process induced by nascent atomic hydrogen and hence aggravated by CP. It is exemplified by the sulfide stress cracking (SSC) of hardened steels in "sour" (i.e., hydrogen sulfide-bearing) service; and

(3) *Liquid Metal Cracking* (LMC). This is the result of liquid or molten metal on a stressed metal (e.g., the effect of mercury on high-strength copper alloys). It sometimes is also called liquid metal embrittlement (LME).

All three of these types of environmental cracking meet the definition provided in the lead paragraph, while corrosion fatigue does not. As described further in Chapter 8.2, chemical cracking of plastics is known as Environmental Stress Cracking (ESC).

8.1.1   Stress Corrosion Cracking

SCC is an electrochemical process and usually occurs in only mildly corrosive environments. A highly stressed area, not necessarily the point of maximum stress, becomes anodic to the adjacent metal. The large cathode-to-anode area ratio causes rapid penetration along the stressed areas, which may comprise either grain boundaries or transgranular slip-planes. The anodic dissolution, together with the tensile stresses which tend to pull the metal apart, causes rapid cracking.

The stresses must be tensile in nature. (*Compressive* stresses, such as the result from shot-peening, are actually used to combat SCC.) The tensile stresses may be residual in nature (as from cold-work, restraint during welding, or uneven cooling) or applied, as from mechanical loading. A common example is the stress due to differential thermal expansion, as of type 300 series stainless steel tubes in a steel-shelled exchange. Tube-bending and tube-rolling are a common source of stress in heat exchangers, while forming and welding are the source of stresses in vessels.

The principal factors involved in SCC are the magnitude of the tensile stress, the nature of the environment, the characteristics of the material, and the length of time or duration of the exposure. These factors interact, and their relative importance varies with the conditions of exposure. The rapidity of failure may vary from a few hours (or even minutes, in extreme cases) to periods of many years.

The most effective means of combatting this type of attack is to select materials of construction which are not susceptible to SCC in the environmental conditions anticipated. This is not always possible. Economic considerations may dictate the use of a susceptible material, or one may not be able to anticipate the presence of agents specific for SCC (especially where they need be present only in trace amounts).

A practical means of alleviating, if not preventing, this type of corrosion is thermal stress relief. The item of equipment is heated to a suitable temperature, held for approximately one hour per-inch of thickness (but usually not less than two hours), and *slowly* cooled. This will normally double or treble the life of equipment. It usually will not prevent SCC altogether, because it is impossible to eliminate stresses completely from operating equipment.

Mechanical stress relief, as done by controlled shot-peening, is also a useful option, especially for rotating equipment which may not be amenable to thermal stress-relief because of distortion effects.

8.1.2   Hydrogen-Assisted Cracking

Hydrogen-assisted cracking (HAC) is caused by access of atomic hydrogen to a susceptible metal surface. Most often, this is caused by the corrosion process itself, in the presence of some agent which prevents the oxidation or dimerization of the nascent atomic hydrogen. In practice, susceptibility is most often associated with the internal stresses of a hardened steel. The most common agent, which is found under anaerobic conditions and is also specific for poisoning the dimerization of hydrogen, is hydrogen sulfide. However, very hard and highly stressed alloy steels can suffer HAC even in salt air, simply from hydrogen generated at the cathodic sites. Characteristically, unlike SCC, the danger of HAC or SSC diminishes rapidly with only moderate increases of temperature, e.g., above 80°C (175°F).

HAC is aggravated by any situation which tends to promote the ingress of atomic hydrogen, as in CP, or when the susceptible material might serve as a cathode in a galvanic couple.

For steels particularly, the most common form of control is to restrict the hardness of equipment, e.g., to below HRC 22, depending upon the specific alloy. NACE Standard MR0175[1] provides the accepted criteria for equipment in sulfide service.

8.1.3   Liquid Metal Cracking

Liquid metal cracking (LMC), also called liquid metal embrittlement (LME), is environmental cracking caused by penetration of molten metal along the grain boundaries of a metal or alloy under tensile strength. One common example is the effect of mercury on brass or other highly stressed or high-strength copper alloys. In this case, even mercury salts in aqueous solution are inimical because of cathodic reduction of the mercury ions to metallic mercury. A mercuric nitrate solution can be used to check for residual stresses in yellow brass. Mercury vapors (such as that which may arise from broken thermometers or blown manometers) have caused failure of silicon bronze and aluminum bronze process equipment.

At elevated temperatures, molten lead, cadmium, and zinc have caused problems with process equipment, especially in austenitic stainless steel construction. Molten copper will cause LMC of steel (e.g., as in welding of copper-clad steel vessels). A nickel sulfide eutectic will cause LMC of nickel-based alloys if there is sulfur contamination of weldments, for example.

8.1.4 Specific Materials

The susceptibility of specific alloy groupings to the three types of environmental cracking are shown in Table 8.1 and discussed below.

Magnesium Alloys

Magnesium alloys which contain aluminum or zinc to improve mechanical strength for aircraft construction may be subject to SCC in atmospheric exposures. The cracking is predominantly transcrystalline. A solution of sodium chloride containing sodium chromate is used as a quality assurance test for alloy evaluation, and is said to correlate well with atmospheric exposures. Being anodic to other engineering metals and alloys, magnesium is not subject to HAC. I am not aware of any reported LMC, but mercury could be expected to cause problems, if encountered.

Aluminum Alloys

Aluminum alloys containing more than 6% magnesium or 12% zinc are subject to SCC in both atmospheric and water exposures, although low alloys and commercially pure aluminum are resistant. Cracking is usually inter-granular and has been studied extensively. Salt solutions containing oxidants are specific for this type of attack, and are used as quality assurance tests. HAC is not

Table 8.1

Susceptibility to Environmental Cracking

Material	SCC	HAC	LMC
Magnesium	Cl^- + CrO_4^{-2}	—	Na, Zn
Aluminum	Cl^- + Ox. Agnts	—	Hg, Na, Sn, Zn
Steel			
Soft	OH^-, NO_3^-, CN^- CO_3^{-2}, NH_3, CO–CO_2–H_2O	—	Cd, Cu, Pb, Sn, Zn
Hard	OH^-, Cl^-	$H°$, H_2S,	Cd, Cu, Pb, Sn, Zn
Stainless			
Martensitic	OH^-	$H°$, H_2S, Cl^-	?
Ferritic	OH^-, Cl^-	—	?
Austenitic	OH^-, Cl^-	—	Cd, Cu, Pb, Zn
Aust., sens.[A]	$S_2O_6^-$, O_2–H_2O	—	?
Copper Alloys	NH_3, NO_3^-, steam	$H°$	Hg
Nickel Alloys			
Alloy 200	OH^-	—	S
Alloy 400	OH^-, HF + O_2	H_2S	S
Alloy 600	OH^-	—	S
Alloy 625	OH^-	H_2S–$Cl^{[B]}$	S
Alloy C-276	OH^-	H_2S–$Cl^{[B]}$	S
Titanium	HNO_3, Cl^- (300°C)	Methanol	Cd, Pb, Sn, Zn

[A] Sensitized.
[B] Severely cold-worked and functioning as a cathode in a galvanic couple.

reported. LMC is probably possible, but mercury contamination usually leads to amalgamation and general corrosion (e.g., in acetic acid storage tanks).

Steel

Steel is subject to SCC, HAC (when hardened), and LMC. The most common SCC agent is caustic, the attack being erroneously named *caustic embrittlement* (see Chapter 27). Steels also suffer SCC on exposure to nitrates, concentrated nitric acid, dilute nitric acid plus manganese dioxide, and by mixtures of carbon monoxide and carbon dioxide with water at elevated temperatures.

At hardnesses above HRC 22, steel suffers HAC by sour environments (i.e., SCC), cyanide, hydrofluoric acid, and thiocyanate solutions.

LMC has occurred due to exposure to molten copper, as in brazing operations and in welding copper-clad equipment.

Stainless Steels

Stainless steels are susceptible to all three types of environmental cracking, contingent upon their type and condition. SCC is encountered with martensitic, ferritic, and austenitic grades in hot caustic. SCC of the type 300 series austenitic grades has been reported in a variety of environments, from foodstuffs to industrial waters and chemicals. In most instances, this is due to the ubiquitous chloride ion. Usually, the SCC is transgranular and multi-branched for type 300 series in both chloride and caustic environments, but may be intergranular if the metal structure is sensitized.

SCC of type 300 series stainless steels is very sensitive to chloride ion concentration, pH, and temperature. High chloride concentrations and low pH are a deadly combination. These stainless steels can suffer Cl^-SCC in a matter of hours in boiling 42 to 45% magnesium chloride solutions.

Heat transfer is an important variable. Stainless steel will often tolerate simple immersion in hot water, yet suffer SCC if used to cool another stream with the same water. SCC of type 300 series condenser tubes is a common experience. A rule of thumb in condenser design is to permit not more than a 50°C (122°F) *tube-wall* temperature. Note that vapor spaces, where salts and deposits can accumulate, are very dangerous. Alternate exposure to steam and water is particularly conducive to SCC.

SCC is electrochemical in nature. Complete deaeration of water (which removes the cathodic depolarizer, oxygen) has been effective in some applications (e.g., for type 300 series waste heat boilers). CP has been useful in such applications as sacrificial metallic coatings on stainless tube inlets. Industrial-type potentiostats are now being employed in some processes in an attempt to maintain the materials at a non-susceptible level of potential.

External stress corrosion cracking (ESCC) of type 300 series stainless steel equipment is occasioned by chloride contamination of the surface (e.g., under insulation). The chlorides may derive from the insulation itself; may be deposited from the atmosphere; or may occur by spillage of water or aqueous solutions. A practical solution is to *paint* the exterior with a chloride- and zinc-free paint system (e.g., a modified silicone or epoxy).

Stainless steels with a higher nickel content (e.g., alloy 254 SMO[†] [S31254], alloy 825 [N08825], alloy 20Cb-3[†] [N08020]) are resistant to chloride SCC but not to cracking by hot caustic. A mixed austenitic-ferrite structure (e.g., type 315 [S31500], alloy 2205 [S31803], alloy 255 [S32550]) is helpful against chlorides, but not entirely reliable. A thorough discussion of how to prevent environmental cracking is covered in MTI Publication No. 15, *Guidelines for Preventing Stress Corrosion Cracking in the Chemical Process Industries.*[2]

The martensitic grades can suffer HAC in the hardened condition. In addition to sour service, salt-laden atmospheres can crack 14% chromium grades, as was observed in attempts to use these steels in lighter-than-air craft.

LMC is occasioned mostly by zinc contamination (e.g., attempts to weld galvanized steel to stainless; and contamination by zinc vapors from galvanized hardware during fire). Overspray from painting adjacent equipment with zinc-pigmented paints has caused LMC during subsequent welding operations. However, lead, bismuth, and cadmium are also possible hazards respective to LMC.

Lead

Lead and its alloys are susceptible to SCC in lead acetate solutions. HAC and LMC are not practical problems.

Copper

Copper and its alloys, when hardened by cold-work, are susceptible to SCC (even pure copper). The high strength alloys, like silicon and aluminum bronze, are notorious in this regard. Brasses and bronzes are particularly subject to SCC by ammonia (i.e., the so-called "season cracking") and by steam. Nitrates and nitrites can also cause cracking, although this may entail cathodic reduction to ammonium ions.

[†] Trade name

I know of only one reported instance of HAC, which was due to the action of an impressed current CP system on a naval bronze ship propeller.

LMC of copper alloys is usually due to mercury contamination.

Nickel

Nickel and its alloys are quite resistant to environmental cracking. However, even nickel (N02200) itself can suffer SCC in concentrated caustic at temperatures of the order of 300°C (575°F). Alloy 400 (Ni-Cu, N04400) suffers SCC by hydrogen fluoride vapors in the presence of air (due to formation of cupric fluoride), as well as in hot caustic and caustic-contaminated steam.

HAC was encountered in a severely cold-worked alloy 400 expansion joint upon exposure to a sulfide-contaminated alkaline steam condensate, under the simultaneous influence of a galvanic couple with a steel liner.

LMC of alloy 400 by mercury vapors has been reported.

Titanium

Titanium suffers environmental cracking in red fuming nitric acid and in anhydrous alcohols. In the latter case, halide contamination will further aggravate the situation.

Zirconium

Zirconium may suffer environmental cracking in the presence of iodine.

8.2 Plastics

Both thermoplastic and thermosetting resin materials can suffer cracking from mechanical, chemical, and ultraviolet effects. Cracking due to the combined action of stress and chemical attack, environmental stress cracking (ESC), has the same specificity as described for metals and alloys, although it is not, of course, electrochemical as in the case of plastic materials.

The polycarbonates are notoriously susceptible, failing by ESC even in atmospheric exposure in industrial areas. The polyolefins are cracked by detergents, wetting agents, and organic solvents (notably ketones and esters). ESC of polypropylene has been reported in silicone oils. Polysulfones may crack even in vapor exposures of ketones and esters. Nylon can suffer ESC in both acidic and alkaline solutions. Acrylonitrile butadiene styrene (ABS) and polystyrene plastics are also reportedly susceptible in unspecified organic solvents.

References

1. NACE Standard MR0175, "Sulfide Stress Cracking Resistant Metallic Materials for Oilfield Equipment" (Houston, TX: NACE International).

2. D. R. McIntyre, C. P. Dillon, *Guidelines for Preventing Stress Corrosion Cracking in the Chemical Process Industries*, MTI Publication No. 15 (Houston, TX: NACE International, 1985).

9

Corrosion Testing

There are many reasons for running laboratory and/or field corrosion tests and many considerations in their design and utilization. The purpose of the test must be clearly defined and understood. In some cases, *only* laboratory tests will suffice (e.g., quality assurance tests), while in others a test under service conditions is essential.

In the formulation of a corrosion test program, one must be aware of the metallurgical and corrosion characteristics of the metals and alloys of interest and of the strengths and weaknesses of nonmetallic materials.

9.1   Material Factors

9.1.1   Composition

The specific chemical composition of a material is a major factor in determining its corrosion behavior. In scientific research, it may be necessary to know the exact composition. In such cases, it is often desirable to obtain an independent analysis. However, in the case of metals and alloys, the composition should be available from the manufacturer. The *heat number* will identify the manufacturing lot, from which record the manufacturer can provide the exact chemical composition.

More often, in ordinary engineering practice, it is sufficient to know the *generic composition* (i.e., if the material is type 304, type 316, etc.) Normally, minor variations *within commercial composition limits* are not the cause of aberrations in service performance (although this is the first thing the novice tends to look for; e.g., is the alloy on the low side of the commercial limit for chromium or molybdenum, for instance?

9.1.2 Homogeneity

This characteristic cannot be quantitatively described, but relates to the uniformity of composition across a plane surface. It is important to know whether a metal contains more than one phase (which might be subject to preferential attack). In cast stainless steel, for example, the amount of ferrite in the otherwise austenitic structure can have an important influence on corrosion. Ferrite is beneficial against chloride pitting, crevice corrosion, and chloride SCC, but can be selectively attacked in reducing acids. Ferrite is indicated by a detectable magnetic permeability, and can be estimated quantitatively. However, ferrite can be converted by heat treatment to the nonmagnetic *sigma phase*, which still can affect adversely corrosion behavior (e.g., in oxidizing acids).

In wrought stainless steels, sensitization of the weld heat-affected zone (HAZ) results in a non-homogeneous structure susceptible to interganular attack (IGA) but undetectable by conventional chemical analysis. Nonmetallic inclusions in metals and extraneous materials in nonmetals can be focal points for corrosion damage.

9.1.3 Stress

Residual stress (e.g., as from cold-work) can affect general corrosion. It may be beneficial, as in the performance of cold-worked copper in vinegar service. More often, it is somewhat harmful (e.g., as indicated by the reaction to acid etching of stamped engine block serial numbers which have been filed or ground off). Residual stress is, of course, very detrimental in services conducive to environmental cracking. Except for the stamped identification numbers on coupons, there should be no substantial residual stress involved in any corrosion test, unless one is studying environmental cracking. In such tests, the *applied* stress should be known at least in a semi-quantitative manner. A number of designs have been developed for stressed coupons, such as U-bends, C-rings, bent beam, tuning forks, etc.

9.1.4 Thermal History

The heat treatment of a metal or alloy should be known (or re-established), as thermal history may have a profound effect on corrosion.

Annealed, normalized, stabilized, sensitized, or stress-relieved items may behave quite differently, depending upon the material and the environment.

Welding is a special case in point. The cast structure of a weld may have a different corrosion resistance than the wrought parent metal (either better or worse). The heat of welding may alter the resistance of the HAZs in the parent metal, while microsegregation in the weld itself may result in diminished corrosion resistance. Normally, the *kind* of welding (if properly performed) does not have a great influence, although there are exceptions.

Usually, testing of welded *coupons* is not recommended to determine resistance of austenitic stainless alloys to weld decay, because the thermal effects can neither be quantified nor reliably reproduced with various thicknesses of metals, number of weld passes, etc. However, welded coupons have been standardized and are useful for many alloy compositions and for evaluation of overmatching (i.e., higher alloy) weldments on a lower alloy.

9.2   Materials Characteristics

The characteristics of metals are discussed in Chapters 10 through 18. However, the following is a list, for easy reference, of specific phenomena associated with different classes of materials.

Aluminum alloys—General corrosion, pitting, IGA, exfoliation, SCC.

Cast Iron—General corrosion, graphitic corrosion.

Steels—General corrosion, localized corrosion, SCC, HAZ effects, weld attack.

Stainless steels—General corrosion, pitting, IGA, SCC, weld attack.

Copper alloys—General corrosion, SCC, dealloying effects.

Nickel alloys—General corrosion, IGA, SCC.

Reactive metals—General corrosion, IGA, SCC, hydrogen attack.

Rubber and elastomers—Softening, hardening, swelling, embrittlement.

Plastics—Softening, hardening, swelling, embrittlement.

Paints and coatings—Softening, swelling, crazing, loss of adhesion.

9.3   Laboratory Tests

There are basically three types of laboratory tests, relating to specific phenomena, quality assurance, and service behavior, respectively.

9.3.1   Corrosion Phenomena Tests

Phenomena-related tests are intended to evaluate the relative resistance of metals and alloys (or of nonmetallic materials) to specific types of attack, e.g., SCC, IGA, pitting, or dealloying. In some cases, there are available standardized environments, coupons, or methodology. These are used for alloy development and studies of materials-related variables. (See also Chapter 9.3.3.)

9.3.2   Quality Assurance Tests

Quality assurance tests are standardized tests which are intended to give a quantitative evaluation of the resistance of metals and alloys to phenomena to which they are known to be susceptible, unless properly formulated and heat-treated. Although manufacturers also perform such tests, it is more often the user who is evaluating the composition, homogeneity, and/or thermal history of a specific lot of material. This may be critical for materials to be used in aggressive chemical or petrochemical processes.

There are standard laboratory tests for dezincification of brass, exfoliation of aluminum, environmental cracking of a number of materials, and particularly for IGA of austenitic stainless steels.

The tests for IGA of austenitic stainless steels are exemplified by ASTM A 262.[1] This specification comprises a screening test (Practice A, Electrolytic Oxalic Acid Etch) and three other total immersion tests (Practices B, C, and D) based on boiling 65% nitric acid, 10% nitric-3% hydrofluoric, and 50% sulfuric ferric sulfate, respectively. Similar tests have been developed for the high-performance stainless steels and nickel-rich and nickel-based alloys.

9.3.3   Service Tests

Service-related laboratory tests may be divided into three categories:

(1)   Laboratory tests may be run in a number of pre-selected standard solutions, in order to categorize new materials by direct comparison with older materials of known behavior;

(2)   The response of materials may be studied regarding their probable behavior under conditions conducive to some specific problem (e.g., velocity effects, heat-transfer behavior) or phenomena (e.g., IGA, SCC); and

(3)   Predictive tests are those which attempt to establish the probable behavior of a material under known conditions, compared to other materials. For example, a test might be run to determine whether type 316L (UNS S31603) would probably be better than type 304L in a particular service.

Note: It is dangerous to try to *accelerate* a test by changing temperatures, velocities, or other conditions, because such changes may drastically alter the corrosion characteristics of the environment (e.g., boiling solutions become anaerobic).

9.3.4   Methodology

A typical laboratory test apparatus and details concerning recommended procedures are found in the NACE Standard TM0169[2] or its ASTM counterpart, ASTM G 31.[3] A laboratory test may be of relatively short duration (e.g., from less than 24 hours and up to several weeks), because small coupons are used, which can be weighed very accurately. A small change in weight (e.g., a few tenths of a milligram) can be detected, which can be calculated as corrosion rate from the appropriate formula.

There are also electrochemical techniques which can be used to evaluate metals or phenomena, based on current/voltage relationships. Some techniques measure the corrosion current developed by small voltage changes, from which the corrosion rate is measured directly. Other techniques develop curves of the current flow caused by incremental changes in an applied voltage. The configuration of the curves indicates active/passive behavior and the corrosion potentials related to pitting, crevice corrosion, and SCC. Such tests should be conducted and evaluated only by trained professionals.

9.4 Field Tests

There are a number of reasons for running field or in-plant corrosion tests in addition to (or in preference to) laboratory tests. It may not be possible to duplicate plant conditions in the laboratory, or very difficult to do so because of temperatures, pressures, flows, contaminants, etc. A field test is more reliable for comparing contemplated alternative materials with existing materials. Field tests are the best and often the only way to monitor the effects of process variables or changes.

There are pitfalls even in field-testing. In addition to the materials-related factors discussed above, the test may not faithfully reproduce such factors as crevices, stresses, or weld-related phenomena. Also, even the field test can only imperfectly evaluate mechanical phenomena, localized corrosion, environmental cracking, heat-transfer effects, and intermittent process contamination. Nevertheless, there are a number of valuable and proven field-testing techniques, as discussed below.

9.4.1 Corrosion Racks

These are devices intended to hold corrosion coupons within process vessels or piping, usually in electrical isolation to prevent galvanic effects. Details can be found in ASTM G 4.[4] A brief description of some of the devices follows:

Bird-Cage

The bird-cage consists of major end-pieces and support rods, within which is enclosed another end-threaded rod carrying the corrosion coupons. Such a device can test a large number of coupons at one time. The disadvantage is that process equipment must be down and open (and otherwise prepared for entry) in order to install or remove the racks. The International Nickel Company formerly supplied such racks, and obtained considerable in-plant exposure data.

Insert Racks

A more modern device consists of a rod/coupon assembly mounted on a welding disk and designed to be supported within an unused nozzle, with the coupons projecting into the process stream. While the equipment must be out of service for installation and removal, the process only entails removal and replacement of a small flange, typically 1½ to 2 in. (3.75 to 5.0 cm).

Dutchman

A *dutchman* is a disk which fits between two flanges (e.g., in a line or between a pipe-vessel connection), with the disk perforated so that it does not impede flow. The dutchman normally is used for somewhat larger flanges than the insert rack, usually with the coupons mounted sideways on a cross-strip.

Slip-In

Slip-in racks are designed for insertion in and removal from operating equipment without shutdown. They include a nozzle recess/packing gland arrangement so that the rod and coupons can be slid in through a full-port gate valve. For services above 2 MPa (300 psi), special high-pressure access devices are available commercially.

9.4.2 Corrosion Devices

These are corrosion-detection or corrosion-measuring devices which do not employ coupons. Among these devices are:

Bayonet Heat Exchangers

These are single-tube exchangers used to evaluate hot-wall or condensation effects within process equipment. (A metal transferring heat to or from an environment may corrode differently than the same metal in a simple immersion situation at the same temperature.)

Hydrogen Probes

Evolution of hydrogen is one indication of corrosion. Hydrogen probes vary from simple single-tube capillary devices (in which diffusion of atomic hydrogen results in dimerization and development of pressure, measured by a pressure gauge) to more sophisticated devices which detect hydrogen on the outside of equipment by electrochemical techniques.

Electrical Resistance (ER) Probes

This technique utilizes a Wheatstone bridge circuitry to measure the corrosion rate through the increased electrical resistance over time of a single, exposed corroding element. The advantage of these devices lies in the ability to detect corrosion changes in a matter of hours, rather than obtaining the average change over a prolonged period, as with coupons. False readings are obtained if conductive films form (e.g., magnetite, iron sulfides, copper plating). The devices are effective in both liquid exposures (conductive or otherwise) and vapor phases.

Electrochemical Devices

In its crudest form, the electrochemical device is simply a galvanic couple, such as an alloy valve in, but electrically isolated from, a steel line connected through either a voltmeter or ammeter. Significant changes in the observed value of potential or current indicates changes in corrosivity (e.g., acid ingress, a large change in pH, or process changes).

In a more sophisticated form such as the polarization admittance instantaneous rate (PAIR) probes, also called linear polarization resistance (LPR) probes, minute voltage changes between paired electrodes measured against a working electrode result in current flow from which *instantaneous* corrosion rates can be obtained. These devices require an electrolyte as the corrosive medium and do not work in the vapor phase. Also, oxidation/reduction reactions other than the corrosion process can yield misleading results. A three-element probe may be used to correct for the conductivity of the medium.

Sidestream Apparatus

This is a device permitting diversion of a small side-stream from a process to an apparatus in which corrosion racks or devices can appraise the result of process changes, inhibition, etc., without disturbing the process itself.

Experimental Equipment

Obviously, experimental installations of pipe sections, valves, pumps, heat exchanger tubes (or even whole tube-bundles) may be made. While these are the most reliable indicators of material performance, the logistics and planning for installation, removal, and evaluation can be difficult. Operations, maintenance, and inspection personnel need to be completely dedicated to the effective operation of such experimental installations if useful data are to be obtained.

9.5 Corrosion Coupons

Both laboratory and field corrosion tests utilize corrosion coupons, sometimes of a standard size. A convenient form is a 1½ in. (38 mm) round coupon, approximately 0.125 in. (3 mm) thick, having a 7/16 in. (11 mm) center hole. These coupons will pass through a 45/50 ground-glass neck joint on a laboratory flask and will fit inside a nominal 1½ in. (38 mm) piping tee or nozzle. Polytetrafluorethylene (PTFE) or other nonconducting spacers are used to electrically isolate the coupons from each other, the rack and the equipment.

9.6 Coupon Evaluation

Corrosion coupons are carefully measured and weighed before exposure. After exposure, they are cleaned, dried, and re-weighed (see NACE TM0169,[2] ASTM G 31,[3] and ASTM G 4[4]), and the corrosion rates calculated:

$$\frac{\text{Penetration}}{\text{mm/y}^{(A)}} = \frac{87.43 \times \text{weight loss (mg)}}{\text{Area (cm}^2) \times \text{sp.gr.} \times \text{hours}} \quad (1)$$

[A] mm/y x 39.37 = mpy (mil per year).

One should also record special observations concerning pitting, crevice corrosion (e.g., under the spacer), SCC, dealloying, etc.

References

1. ASTM Standard A 262, "Standard Practices for Detecting Susceptibility to Intergranular Attack in Austenitic Stainless Steels" (Philadelphia, PA: ASTM).

2. NACE Standard TM 0169, "Laboratory Corrosion Testing of Metals for the Process Industries" (Houston, TX: NACE International).

3. ASTM Standard G 31, "Recommended Practice for Laboratory Immersion Corrosion Testing of Metals" (Philadelphia, PA: ASTM).

4. ASTM Standard G 4, "Method for Conducting Plant Corrosion Coupon Tests in Plant Equipment" (Philadelphia, PA: ASTM).

Suggested Resource Information

H. Hack, *Corroson Testing Made Easy–Galvanic Corrosion* (Houston, TX: NACE International, 1994).

B. J. Moniz, W. I. Pollock, eds., *Process Industries Corrosion–The Theory and Practice* (Houston, TX: NACE International, 1988), pp. 67, 85, and 123.

A. J. Sedriks, *Corrosion Testing Made Easy–Stress Corrosion Cracking Test Methods* (Houston, TX: NACE International, 1990).

E. Verink, *Corrosion Testing Made Easy* (Houston, TX: NACE International, 1994).

SECTION III

Materials

10

Light Metals

10.1 Structure of Metals

The following comments are generally applicable to all metals and alloys and serve as a common introduction to the subject.

The atomic structure of an element consists of a nucleus containing neutral particles (neutrons) and positively charged particles (protons), surrounded by orbiting negative particles (*electrons*). (See Figure 10.1.) It is the potential loss of electrons that accounts for the electrochemical nature of corrosion.

A small orderly arrangement of atoms in a geometric configuration constitutes a *unit cell* (Figure 10.2), while an orderly stacking of unit cells forms a metallic crystal, or grain; the terms are synonymous.

In outward appearance, a metal or alloy is a homogeneous solid. Under the microscope, however, it is readily observed to consist of many individual crystals. Unless it is extremely pure, it also will contain constituents of different compositions (e.g., intermetallic compounds in certain aluminum alloys, different phases in steel, free graphite in cast iron).

10.2 Magnesium

The two outstanding characteristics of magnesium are its very light weight (the density of approximately 1.8 being approximately two-thirds that of aluminum) and its highly anodic position in the galvanic series in water. It is anodic to all common engineering metals in aqueous environments.

Magnesium alloys may contain small amounts of aluminum, zinc, or tin. As a material of construction, magnesium alloys are used primarily for aircraft (e.g., engine parts, nose pieces, landing

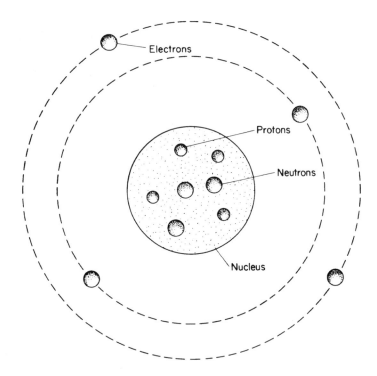

Figure 10.1 Atomic structure.

wheels, parts of the fuselage, oil and gas tanks), for moving parts in machinery (e.g., blowers), and in light-weight portable structures (e.g., ladders). For even mildly corrosive service, magnesium alloys may require anodizing (to reinforce the surface film) or painting, although magnesium drums and tanks are used for highly

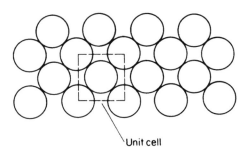

Figure 10.2 Unit cells.

specific services, such as phenol and methyl bromide. Surprisingly, it has good resistance to hydrofluoric acid, because of the insoluble corrosion product film.

Magnesium is widely used as sacrificial anodes in the CP of steel piping or structures underground. Aside from the aircraft industry, magnesium is rarely encountered by the engineer concerned with materials of construction.

10.3 Aluminum and Its Alloys

10.3.1 General

Aluminum and its alloys are light-weight (density approximately 2.7 g/cm^3, approximately one-third that of carbon steel), reasonably strong, and quite resistant to natural environments in spite of their anodic position in the galvanic series. (This resistance is due to the presence of a protective oxide film.) These materials are relatively easy to fabricate, available in most standard forms, and can be joined by most of the common methods of welding and brazing.

Commercially pure aluminum and its alloys have UNS numbers (see Chapter 3.4.5), with identifications derived from the older Aluminum Association (AA) designations. The general format is A9xxxx for wrought materials, and A0xxxx for castings. As seen in Table 10.1, the second digit (A9xxxx) indicates the major alloying element. The third digit (A9xxxx) indicates whether there are controls on impurities, while the last two digits (A9xxxx) are arbitrary holdovers from the older AA system. Alloys commonly used in the process industries include A93003, A95083, A95086, A95154, and A96061.

10.3.2 Properties

Annealed aluminum alloys (e.g., A91100) have mechanical properties of the order of 110 MPa (16 ksi) tensile strength, and 41 MPa (6 ksi) yield strength. These alloys can be hardened by cold-work (i.e., *strain-hardened*). Other wrought alloys can be heat-treated (i.e., *age-hardened* or *precipitation-hardened*). In such cases, the alloy designations carry a suffix: "H" for strain-hardened, "T" for tempered (i.e., precipitation-hardened, followed by drawing to the desired hardness). The "T" designation is followed by a number indicative of the specific procedure.

Table 10.1

UNS Numbers for Aluminum Alloys

Alloy No.	Example	Major Alloying Element
A91xxx	A91100	None, 99.00% aluminum minimum
A92xxx	A92020	Copper
A93xxx	A93003	Manganese
A94xxx	A94002	Silicon
A95xxx	A95154	Magnesium
A96xxx	A96063	Magnesium and silicon
A97xxx	A97001	Zinc
A98xxx	A98013	Miscellaneous (e.g., chromium)

In the process industries, one does not look for anything stronger than, for example, A96061-T6. This indicates an alloy of approximately 310 MPa (45,000 psi) tensile, 276 MPa (40,000 psi) yield, and approximately 12% elongation. One of the strongest available heat-treated alloys is A97075 in the T-4 condition, which can attain 572 MPa (83,000 psi) yield strength, while maintaining 10% elongation.

For most purposes, joining with the inert gas-shielded welding processes, e.g., GTAW (TIG) or GMAW (TIG) is preferred, but shielded metal arc (SMA or stick electrode) is acceptable, and gas-welding is permissible. Of course, in hardened alloys, the strength of HAZ is drastically reduced by the annealing effect of the welding heat input.

Aluminum has excellent low-temperature properties with high impact strengths and is used widely in noncorrosive cryogenic services, such as gas-treating and air separation.

10.3.3 Corrosion

Usually, corrosion resistance of aluminum in uncontaminated atmospheres is good. However, it will become unsightly after a time because of the initial rate of attack, which stifles itself by the development of a protective film. For architectural applications, an electrolytically-induced film may be pre-developed by a process called *anodizing*. This gives a more aesthetically pleasing and more stable surface.

The A92000 series of alloys (and the similar copper-bearing castings of the A02000 series) do *not* have good resistance. This is particularly evident in contaminated industrial or marine atmospheres. An insidious subsurface intergranular attack (*exfoliation*) develops, and corrosion proceeds along the grain-boundaries of the cathodic copper-rich phases. The attack may not be evident on the surface, which may show only a frosted appearance, while the thickness of the metal may actually increase as a result of the corrosion products within the alloy structure. Aluminum alloys intended for industrial applications should contain no more than 2.5% copper.

Except for the copper-bearing grades, aluminum has good resistance to many natural waters, as witnessed by the use of aluminum boats even in seawater. However, aluminum is subject to pitting under stagnant conditions in saline waters, particularly as a heat transfer surface (e.g., in heat exchangers).

In water and dilute aqueous solutions, pH from 4.5 to 9.5 should show rates of less than one mpy. The situation is complicated by chlorides or other halides, deposits, and by the characteristic nature of aluminum, which is *amphoteric* (that is subject to attack by either acidic [pH< 4] or alkaline [pH >10] environments). Because aluminum is anodic to most other common metals except zinc and magnesium, it is subject to attack by the ions of heavy metals (e.g., iron, copper, lead, and mercury) due to cementation. Heavy metal ions easily can be present, of course, in industrial waters, as a result from corrosion of equipment upstream.

Many organic chemicals are fully compatible with aluminum. Aldehydes, ketones, esters, amines, and organic acids and anhydrides can be shipped and stored in aluminum, as dry refined products. It should be emphasized, however, that some organic chlorides (i.e., chloro-organic compounds such as chloroform or ethylene dichloride) and alcohols can react catastrophically with aluminum, with the attendant danger of fires or explosion due to hydrogen liberated in the corrosion process. In some solvents, water may be an effective inhibitor. This is not true universally, however, as in the case of hydrolyzable organic chlorides (which simply liberate hydrochloric acid upon warming with water).

Among the inorganic products, concentrated nitric acid (greater than 90%) and hydrogen peroxide are commercially handled in aluminum. The non-oxidizing acids, and alkalis or *aqueous* ammonia derivatives, are corrosive.

Even in relatively innocuous media, galvanic couples (except with zinc) must be avoided. Galvanized surfaces and zinc-pigmented coatings afford CP to aluminum. CP is also utilized in Alclad[†] products, in which a surface layer of an anodic aluminum alloy corrodes sacrificially to prevent pitting of a more cathodic substrate.

In addition to the exfoliation previously described, some aluminum alloys (e.g., A97074) are subject to IGA of the weld HAZ in corrosive environments.

Environmental cracking is encountered only in the heat-treated, high-strength alloys. Such cracking is a problem primarily in aircraft and marine structures seeking a high strength-to-weight ratio. A typically susceptible material is A97075-T6. However, the same alloy in the T73 temper has good resistance to environmental cracking, which illustrates once again the importance of sound materials engineering.

In the CPI, the predominant aluminum alloys are those of the 3000, 5000, and 6000 series. High-strength alloys of the 7000 series are subject to SCC.

Suggested Resource Information

ASM Handbook, Vol. 2, *Properties and Selection: Nonferrous Alloys and Special-Purpose Materials* (Materials Park, OH: ASM International, 1991).

Aluminum and Aluminum Alloys (Materials Park, OH: ASM International, 1993).

Aluminum in Food and Chemicals (Washington, DC: The Aluminum Association).

The Metals Red Book, Vol. 2, *Nonferrous Metals* (Edmonton, Alberta, CAN: CASTI Publishing, 1993).

B. J. Moniz, W. I. Pollock, eds., *Process Industries Corrosion–The Theory and Practice* (Houston, TX: NACE International, 1988), p. 551.

I. J. Polmear, *Light Alloys* (London, UK: Edward Arnold, 1989).

11

Iron and Steel

Commercial cast irons and steels are basic materials of construction for a variety of industries. We are not concerned with "pig" or ingot iron, nor with the now obsolete wrought iron (although it was once commercially important). The relatively pure low-carbon irons, such as carbonyl iron and electrolytic iron are not structural materials.

11.1 Cast Irons

Various types of cast irons are used widely, especially for pipes, valves, pumps, and certain mechanical parts. Cast iron is an alloy of iron, silicon, and carbon. The carbon content varies from approximately 1.7 to 4.5%, most of which is present in insoluble form (e.g., graphite) in this range. This definition adequately covers the unalloyed gray irons, white irons, malleable irons and ductile irons.

11.1.1 Unalloyed Cast Irons

The most common form of unalloyed cast irons is *gray cast iron*, although the process industries today use ductile cast iron (DCI) almost exclusively (see further below). When the casting is allowed to cool slowly, the insoluble carbon precipitates as flakes of graphite, which are the outstanding feature of this material and cause its typically brittle behavior (because the material fractures along the graphite flakes). Gray irons are quite soft and readily machinable. A typical microstructure is shown in Figure 11.1.

White iron is produced by rapid cooling of a gray cast iron of controlled composition, that low silicon, high manganese, to produce an alloy that is hard, brittle, and practically un-machinable. The carbon is retained as a *dissolved* solid, in the form of iron carbides. White iron is used primarily for wear resistance. *Chilled*

Figure 11.1 Microstructure of gray cast iron.

iron is a duplex material, having a wear-resistant white iron surface (produced by rapid surface cooling) over a comparatively tougher gray iron core.

A relatively ductile material called *malleable iron* can be produced by prolonged heat treatment of white iron. A 30-hour treatment at 925°C (1,700°F), followed by an equivalent period of slow cooling, allows the graphite to precipitate as nodules rather than flakes. Ductility is therefore much less impaired, compared to gray cast iron. Malleable irons have been used in fittings, machinery, tools, and automotive parts.

Another method of improving the mechanical properties of cast iron is through the addition of "inoculants." Minor additions of calcium silicide (in the proprietary Meehanite[†] process) or of nickel plus ferrosilicon (the Ni-Tensyl[†] process) cause the graphite to separate as fine flakes, rather than the coarser flakes in gray iron. This substantially improves ductility without significantly changing the chemical composition of the cast iron.

The best modern form of cast iron, having superior mechanical properties and equivalent corrosion resistance, is *ductile cast iron* (DCI). The addition of a small amount of nickel-magnesium alloy to cast iron causes the graphite to precipitate as *spheroids* rather than as flakes (Figure 11.2). This results in a ductility approaching that of steel. Ductile iron can be produced to have as much as 18% elongation, while some wrought carbon steels have no more than 20%. Ductile iron castings also can be produced to have improved low-temperature impact properties (i.e., low ductility transition temperature, NDTT) by control of the phosphorous, silicon, and alloy content, as well as the thermal treatment.

[†] Trade name

Figure 11.2 Microstructure of ductile cast iron.

11.1.2 Alloy Cast Irons

There are several types of commercially important alloy cast irons, in which the alloy additions substantially modify mechanical and physical properties as well as corrosion resistance.

Molybdenum

Molybdenum may be added to improve strength or, along with other alloying elements, to improve corrosion resistance in chloride media.

Silicon

Silicon is added to cast iron in the range from 11 to 14% to produce an alloy with superior resistance to hot sulfuric acid (e.g., *Duriron*[†]). Small amounts of molybdenum or chromium are also sometimes added to improve resistance in the presence of chloride contamination.

Nickel

Nickel is another common alloying element. Nickel is added in amounts varying from 0.5 to 6% in engineering-grade gray irons. At approximately 4.5%, it produces a martensitic gray iron with outstanding resistance to abrasion and wear (e.g., Ni-Hard[†]). Austenitic, nonmagnetic gray irons containing 14 to 38% nickel, such as the several grades of Ni-Resist[†], have outstanding resistance to corrosion and moderately high temperatures, as well as having very low coefficients of thermal expansion.

Copper

Copper is a mild strengthener, increasing resistance to wear and to certain types of corrosion.

Chromium

Chromium is added to cast irons in amounts varying from 0.15 to 1%, in order to improve resistance to graphitic corrosion. In amounts from 1 to 1.5%, it is added to increase high temperature oxidation resistance. In special alloys, up to 35%, chromium may be added for resistance, both to corrosion and high temperature oxidation.

11.2 Steels

Steel is an alloy of iron and carbon, containing small amounts of other alloying elements or residual elements as well. It is the presence of carbon and its effect upon response to heat treatment that changes iron from a laboratory curiosity to an engineering material.

In the manufacture of steel, iron ore is reduced in a blast furnace to produce pig iron. Pig iron contains impurities (e.g., carbon, silicon, phosphorous, sulfur, etc.) which makes it hard and brittle. It must be refined, and the alloy content controlled, in order to obtain suitable properties. A newer method of producing steel is with the basic oxygen furnace, in which pure oxygen (rather than air) is blown through the molten metal. To prevent the reaction of residual oxygen with dissolved carbon during solidification, steel may be "killed" (i.e., made to lie quietly in the mold) by the addition of deoxidants such as silicon or aluminum. Killed steels are used down to $-28.9°C$ $(-20°F)$, at least in thinner sections, because of their improved NDTT as compared with ordinary steels. Permissible temperatures will vary with thickness and limits of $-6°C$ $(21°F)$ are sometimes invoked for vessels in cold-temperature service.

11.2.1 Carbon Steels

Carbon steels are primarily iron and carbon, with small amounts of manganese. They are the workhorse material for structural members, sheet, plate, pipe, and tubing.

Steels that have been worked or wrought while hot will be covered with a black mill-scale (i.e., magnetite, Fe_3O_4) on the surfaces, and are sometimes called *black iron*. *Cold-rolled* steels have a bright surface, accurate cross-section, and increased yield and tensile strength. The latter are preferred for bar-stock to be used for rods, shafts, etc.

11.2.2 Principles of Heat Treatment

Carbon and low-alloy steels occupy an essential place among materials of construction, precisely because of the potential range of hardness, strength, and other mechanical properties. These are achieved primarily through heat treatment.

Iron has three allotropic crystal forms (i.e., alpha, gamma, and delta) which exist at different specific temperatures from room temperature up to the melting point, and have different capacities for dissolving carbon.

A *phase diagram* (Figure 11.3) best illustrates the following discussion. Of primary concern is the alpha (ferrite: body-centered) and gamma (austenite: face-centered) crystal forms. Ferrite converts to austenite in the vicinity of 910°C (1,670°F).

The hardening of steel is due to a combination of the allotropic transformation and the different solubility of carbon in the two crystalline forms of iron. At room temperature ferrite, carbon is

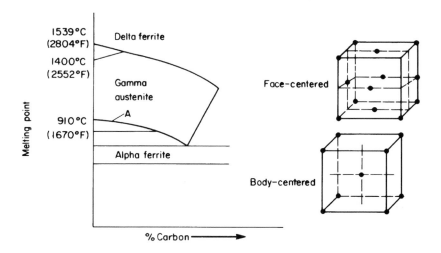

Figure 11.3 Simplified iron-carbon phase diagram.

soluble only to approximately 0.008%, any excess being in the form of iron carbides (i.e., cementite). On the other hand, the austenite which is formed at high temperatures can dissolve up to approximately 2% carbon.

A typical carbon steel might contain approximately 0.30% carbon. At room temperature, approximately 0.29% of the carbon is dispersed in the ferrite structure as alternate platelets of cementite. The mixture of ferrite and cementite is called *pearlite*, because it has the appearance under the microscope of mother-of-pearl.

When such a steel is heated to the transformation temperature (point A in Figure 11.3), the austenite phase is formed and *all* of the 0.30% carbon dissolves. If the alloy is slowly cooled, the austenite reverts to ferrite and the pearlite is also reformed. The process of heating and *slow* cooling is called *annealing* (or *normalizing*, at a somewhat lower temperature). There is little, if any, change in mechanical properties under these conditions.

However, if the heated steel in its austenite form is rapidly cooled (i.e., quenched), the reversion to ferrite is very rapid while the precipitation of carbon is much slower. The carbon atoms then become entrapped in the ferrite lattice, stretching and distorting the structure. This gives a distorted, acicular structure called martensite (Figure 11.4), which is very hard and brittle. In this quenched condition, the material is both at maximum hardness and strength and minimum ductility or toughness. In most engineering applications, a combination of toughness and strength is desired, so some compromise must be effected.

This compromise is attained by reheating the hardened steel to some temperature *below* the lower critical temperature of approximately 720°C (1,330°F). This procedure is known as *tempering* or *drawing*, and allows the "logjam" of iron and carbon atoms to sort itself out. The higher the temperature and the longer the time, the more the iron and carbon revert from the martensite to the ferrite-pearlite structure. A quenched and tempered steel will have much higher strength and less ductility than an annealed or normalized steel, but lower strength and more ductility than the same material in the fully hardened condition.

In hardening a steel, the *rate* of cooling is critical. For a given composition, it is easier to completely harden (i.e., "through-harden") a smaller diameter piece than a thicker one. The latter

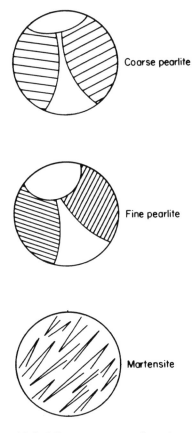

Coarse pearlite

Fine pearlite

Martensite

Figure 11.4 Microstructures of steel.

may retain a softer core than the outside surface, giving a hardness profile across the diameter (Figure 11.5), which may or may not be desirable for the intended end use.

11.2.3 Alloy Steels

Alloying elements, in small amounts, affect primarily the *rate* of cooling required to harden the steel. For example, with small additions of chromium, nickel, and molybdenum, heavier sections can be through-hardened. By the same token, a less drastic quenching medium (e.g., oil instead of water) can be used. More highly alloyed steels can be "quenched" in still air. Also, it should be remembered that, in welding, the weld proper and adjacent areas will be quenched by the mass of cold metal surrounding the joint, causing hardening and leaving high residual stresses.

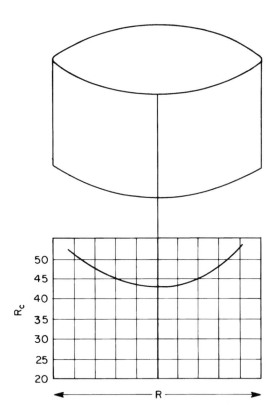

Figure 11.5 Hardness profile of steel bar.

As a general rule, the dividing line between low-alloy and high-alloy steels is approximately 5% total alloying elements. Low-alloy steels were developed primarily to control response to heat treatment and extend mechanical properties. Their corrosion resistance is usually not significantly different from that of carbon steel, although some chromium-molybdenum grades have improved resistance to hydrogen effects (Chapter 30) at elevated temperatures and graphitization (Chapter 31). High-alloy steels are more often made for improved corrosion resistance, with the exception of a few specialty steels (e.g., Hadfield's manganese steel for abrasion resistance).

The effects of the individual alloying elements are as follows:

Carbon

Carbon is the principal hardening element. In fact, the percentage of carbon can be estimated from the maximum hardness induced by heat treatment.

Manganese

Manganese is a deoxidizer and desulfurizer which also contributes to strength and hardness. A major purpose of desulfurizing steel is to improve hot-workability, since iron sulfides cause cracking (i.e., hot-shortness) during hot-working. Further, a manganese-to-carbon ratio of not less than 3:1 is beneficial for impact strength and nil ductility transition temperature (NDTT).

Silicon

Silicon is a principal deoxidizer. The amount of retained silicon will vary with deoxidation practices. A "killed" steel may contain as much as 0.6%, although structural steels usually have a range of 0.15 to 0.30% silicon.

Aluminum

Aluminum is used to complete the deoxidation practice; and, silicon-killed and aluminum-finished steels are used at moderately low temperatures.

Phosphorus

Phosphorus is primarily an impurity, decreasing ductility, and toughness. A maximum phosphorus content of 0.04% or less is commonly specified.

Sulfur

Sulfur is likewise undesirable, except where improved machinability is required. Manganese sulfides break up the chips during machining, but they have an adverse affect upon ductility and impact strength. Sulfur content is usually held to 0.05% or less, except in resulfurized free-cutting grades (which may contain from 0.1 to 0.3% sulfur).

Copper

Copper is added only to improve resistance to atmospheric corrosion, which it probably does by scavenging sulfur. Only small concentrations of copper can be tolerated (not more than 0.2 to 0.3%) because of its low solubility, and hot-short or LMC effects.

The *major* alloying elements that affect corrosion as well as metallurgical characteristics and response to heat treatment, are:

Chromium

Chromium is added to increase the depth to which thick sections can be hardened; to provide abrasion resistance; to provide higher hardness of carburized or carbonitrided surfaces; to improve corrosion and oxidation resistance; to improve resistance to high-temperature, high-pressure hydrogen; and to improve resistance to high-temperature graphitization.

Molybdenum

Molybdenum is added to prevent graphitization and give close control of hardenability, while increasing high temperature tensile and creep strength. It is used at approximately 0.5% concentration in conjunction with chromium and also is effective in preventing temper embrittlement.

Nickel

Nickel is added to improve toughness (particularly NDTT), response to heat treatment, and corrosion resistance.

Vanadium

Vanadium is added to refine grain size and improve mechanical properties, as well as to increase the hardenability of medium-carbon steels.

A combination of two or more alloying elements usually imparts some of the characteristic good properties of each. Chromium-nickel steels develop good hardening properties with excellent ductility, while chromium-nickel-molybdenum steels develop even better hardenability with only a slight reduction in ductility. The carbide-forming elements (e.g., chromium, molybdenum, vanadium) also increase resistance to hydrogen attack at elevated temperatures and pressures, under conditions which lead to embrittlement, blistering, or methanation of carbon steels.

11.3 Numbering

The traditional numbering system for steels was developed by Society of Automotive Engineers (SAE). A similar system was that of the American Iron and Steel Institute (AISI). Together with the numbering system for other alloys, these have been assimilated into the UNS, as described in Chapter 3.4.5.

Each steel is assigned an identifying number consisting of a letter (usually "G" or "K" for carbon and low alloy steels) and a five-digit number. The first two digits codify the major alloy additions, and the next two the carbon content (expressed in hundredths of a percent). The final digit encodes any special requirements. For example, a plain carbon steel of 0.20% carbon (SAE [AISI] 1020) is numbered UNS G10200. G31300 is a nickel-chromium steel of 0.30% carbon, while G43xx0 is chromium-nickel-molybdenum, and G61xx0 is a chromium-vanadium steel of a specific carbon content xx.

Suggested Resource Information

ASM Handbook, Vol. 1, *Properties and Selection–Iron, Steels, and High-Performance Alloys* (Materials Park, OH: ASM International, 1990).

R. M. Davison, J. D. Redmond, "Practical Guide to Using Duplex Stainless Steels," *Materials Performance* 21, 1 (1990): p. 57.

G. Krauss, *Principles of Heat Treatment of Steel* (Materials Park, OH: ASM International, 1980).

W. T. Lankford, Jr., N. L. Samways, R. F. Craven, H. E. McGannon, *The Making, Shaping, and Treatment of Steel* (Pittsburgh, PA: Association of Iron and Steel Engineers, 1985); available from ASM International.

Metals and Alloys in the Unified Numbering System, (Warrendale, PA: SAE).

The Metals Black Book, Vol. 1, *Ferrous Metals* (Edmonton, Alberta, CAN: CASTI Publishing, 1992); available from NACE International.

B. J. Moniz, W. I. Pollock, eds., *Process Industries Corrosion–The Theory and Practice* (Houston, TX: NACE International, 1988), p. 373.

12

Stainless Steels

The addition of a minimum of approximately 11% chromium produces a stainless steel. There are many compositions of stainless steel belonging to different "families," from the basic 11 to 13% chromium steel to complex alloys containing chromium, nickel, molybdenum, copper, etc. Stainless steels are used in a variety of applications ranging from simple protection from iron contamination (e.g., of pure water, foodstuffs, or refined chemicals) to the most demanding corrosive chemical services.

12.1   Nature of Stainless Steel

The addition of approximately 12% chromium to steel produces a synthetic "noble metal," with little tendency to react with most natural environments. The difference between stainless steels and the true noble metals of silver, gold, and platinum is that the nobility or *passivity*, as it is commonly called, of the stainless steel, can be induced, reinforced, or removed by chemical or electrochemical reactions.

Passivity is a surface phenomenon, associated primarily with an oxide film or with adsorbed oxygen. Depending upon the stability, reactivity, and solubility of the film in a given environment, the stainless steels may be either passive or active. In the active condition, they may be less resistant than ordinary steel in hot caustic, molten salts, and reducing acids, for example.

The passive protective file on stainless steels is formed by reaction with oxygen (which is why it resists most natural environments) or with oxidizing agents (e.g., nitric acid). The film is removed by reaction with hydrogen or reducing agents. The film is penetrated by some species, notably chloride ions. Stainless steels tend to be rapidly attacked by reducing acids or hot caustic, for example, and pitted (or cracked) by chloride environments, but tend to resist nitric acid, other oxidizing acids, peroxides, etc. The film can also be removed mechanically, as by abrasion, wear, or erosion, but re-forms readily on exposure to air or aerated water.

Passivity is the normal state of existence for stainless steels, and they are usually several tenths of a volt *cathodic* to carbon steel in natural waters and aqueous solutions. Only under special conditions will they become active, in which case they are *anodic* to steel.

Passivation treatments are chemical treatments used both to remove iron contamination and provide a more homogeneous surface to corrosive environments. Typically, dilute nitric acid is used, but hydrogen peroxide, ammonium persulfate, and similar oxidizing solutions may also be employed, as can electrochemical passivation. Such treatment slightly thickens the protective oxide film. Non-oxidizing solutions do not passivate, nor do nitric-hydrofluoric pastes or solutions. However, a simple water rinse after non-passivating chemical cleaning will restore normal passivation.

12.2   Types of Stainless Steel

Chapter 11 discusses the ferrite/austenite/martensite transformation in steel and alloy steels, based on the iron-carbon phase diagram. A similar diagram (Figure 12.1) is used to explain the phase transformations in iron-chromium and related alloys.

When chromium is the only major alloying element, and carbon is held essentially constant at a moderate value, the conditions under which martensitic and ferritic stainless steels are formed is explained by "the gamma loop." (*Note*: γ is the Greek letter used to denote the austenite phase.)

As illustrated in Figure 12.1, an increase in temperature takes the alloy through the gamma loop up to approximately 12% chromium content. Because they undergo the ferrite/austenite transformation, these 11 to 13% stainless alloys can be hardened to a martensitic structure by cooling from above the transformation temperature. Above approximately this chromium content (the next level of alloying is usually around 17%), the alloy never passes through the gamma loop, and therefore, remains ferritic and of its original hardness, regardless of heat treatment.

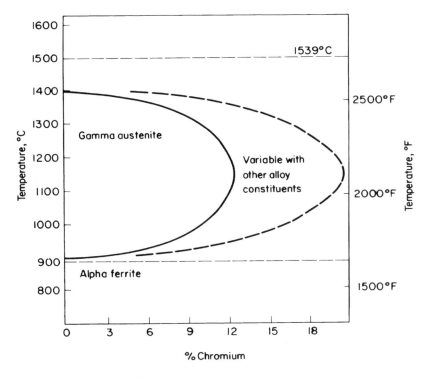

Figure 12.1 The gamma loop.

If the chromium is in the 16 to 19% range and approximately 8 to 14% nickel is added, the phase diagram would show the gamma loop expanded to fill the diagram. Such type 300 series alloys are *austenitic* over the normal range of temperature, rather than ferritic or martensitic.

Nominal compositions of typical wrought stainless steels are given in Table 12.1.

12.2.1 Martensitic Stainless Steels

Because of the combination of hardenability with stainless properties, the martensitic grades are of interest for cutlery, turbine blades, and high temperature parts. Their high order of resistance to atmospheric corrosion, compared with steel, places them into the stainless category. They are, however, of limited usefulness in process exposures compared with the more highly alloyed grades. The martensitic grades are exemplified by types 410 (S41000) and 420 (S42000). These and related alloys are somewhat difficult to

Table 12.1

Wrought Stainless Steels, nominal wt.%

Name	UNS	Cr	Ni	C (max)	Other
Type 410	S41000	11.5	—	0.15	—
Type 430	S43000	16	—	0.12	—
Type 446	S44600	23	—	0.20	—
Type 304	S30400	18	8	0.08	—
Type 304L	S30403	18	8	0.03	—
Type 321	S32100	17	9	0.08	Ti 5 × C min
Type 347	S34700	17	9	0.08	Cb 10 × C min
Type 316	S31600	16	10	0.08	Mo 2
Type 316L	S31603	16	10	0.03	Mo 2
Type 317	S31700	18	11	0.08	Mo 3
Type 317L	S31703	18	11	0.03	Mo 3
Type 309	S30900	22	12	0.20	—
Type 310	S31000	24	19	0.25	—
Alloy 20Cb-3[†]	N08020	19	32	0.07	Mo 2, Cu 3, Cb 8 × C

fabricate by welding, as they require extensive pre- and post-weld heat treatment, because of their air-hardening characteristic. Their transformation rate is so sluggish that a quench in oil or water is not required for hardening, as a slow air cool suffices to effect the martensite transformation.

The greatest weakness of the martensitic grades is a suscepti-bility to absorption of atomic hydrogen, resulting in hydrogen-assisted corrosion (HAC), (see Chapter 8), notably in sulfide environments. Their low-temperature impact resistance is also poor and when cooled rapidly from elevated temperature they are susceptible to IGA (see Chapter 7).

12.2.2 Ferritic Stainless Steels

The ferritic stainless steel class of alloys starts in the 15 to 18% chromium range, although they can go up to approximately 30%. Well outside the gamma loop, they are wholly ferritic. Note that both martensitic and ferritic grades are ferromagnetic, regardless of heat treatment.

The ferritic grades, exemplified by type 430 (S43000), have

[†] Trade name

corrosion resistance superior to the martensitic grades, by virtue primarily of their higher chromium content. They are commonly used for nitric acid services, water, food processing, automobile trim, and architectural applications. Their high temperature oxidation resistance is also good, and they are used for furnace and turbine parts. Their impact resistance and weldability are rather poor, even though they are not hardenable by heat treatment. Welding may result in brittleness and/or diminished corrosion resistance, unless proper pre- and post-weld heat treatment are employed. For example, fast cooling can result in IGA (the reverse of the situation for the austenitic type 304, for example).

Ferritic stainless steels have poor impact resistance at low temperatures, sometimes even at relatively warm temperatures, as discussed further below. They are not susceptible to hydrogen attack, but can suffer SCC in caustic environments, and to a much lesser extent, in chlorides. On heating, as by welding, grain growth can occur, with attendant embrittlement.

Modern controlled-melting and other improved practices have permitted development of a new group of "superferritic stainless steels." These are molybdenum-bearing variants with very low contents of carbon, nitrogen, and other interstitial elements. Such superferritic grades (e.g., alloys 26Cr-1Mo [26-1, S44627], 29-4C[†] [S44735], and Sea-Cure[†] [S44660]) have improved resistance to chloride pitting and SCC and improved general corrosion resistance in some environments, but have some deficiencies in impact strength (e.g., high nil ductility transition temperature [NDTT] even above 40°C [104°F]) and weldability.

12.2.3 Austenitic Stainless Steels

The group, austenitic stainless steels, characteristically non-magnetic, is the most important for process industry applications. By virtue of their austenite-forming alloy additions, notably nickel and manganese, these stainless steels have the face-centered austenite structure from far below zero up to near melting temperatures. They are not hardenable by heat treatment, but can be strain-hardened by cold-work. This also induces a small amount of ferromagnetism. (*Note:* In the cast varieties, a small amount of magnetism is induced deliberately through an imbalance in austenite-forming vs ferrite-forming elements in order to improve casting characteristics.)

The "garden-variety"-type 300 series austenitic stainless steels are exemplified by type 304. Such alloys have a rare combination of corrosion resistance, high-temperature strength and oxidation resistance, ease of fabrication and weldability, and good ductility and impact resistance down to at least $-183°C$ ($-216°F$). Their mechanical properties, in general, are excellent. Improved chemical resistance is obtained by molybdenum additions (e.g., type 316), and resistance to IGA by stabilization (e.g., types 347 with Cb and 321 with Ti), or low carbon content (e.g., types 304L and 316L). Today, both base metal and surface chemistries are improved by argon-oxidation-decarburization (AOD) and other mill practices, and the former differences in mechanical properties between regular- and low-carbon grades largely may be eliminated. Such products may be dual labeled as 304/304L, for example, because they meet all ASTM requirements for both grades due to pin-point control of alloying elements.

The most highly alloyed of this group is type 317L (S31703) with a nominal composition of 18%Cr-11%Ni-3%Mo-0.03%C, which has been used widely in the pulp and paper industry. Modern variants have somewhat higher molybdenum content and are strengthened with nitrogen, e.g., types 317LN (S31753), 317LM (S31725), and 317L4 (S31726). Nitrogen is also an austenitizing constituent which can replace nickel to a certain extent.

When substantial amounts of nickel are replaced by manganese, an analogous group of alloys is formed, exemplified by the 8%Mn-4%Ni alloy (S20200). Other grades can be *precipitation-hardened* (e.g., alloys 17-4 PH [S17400] and PH 15-7 Mo† [S15700]) while retaining corrosion resistance similar to that of type 304 or 316.

Higher austenitic alloys, such as types 309 (22 Cr-12Ni, S30900) and 310 (25Cr-20Ni, S31000), are used for high temperature applications and for welding type 300 series to steel. There are also *duplex* alloys (i.e., mixed ferrite-austenite structure) for improved corrosion resistance, although at some cost in fabricability; an example is type 329 (S32900). (See Chapter 12.2.4.)

12.2.4 High-Performance Stainless Steels

In order to truly meet the definition of a stainless steel, the material must be a ferrous alloy and must contain more than 50% iron. Those special alloys which are strictly neither iron-based nor nickel-based are covered in Chapter 13. Like the type 300 series varieties, the high-performance stainless steels are usually made by the argon-oxygen-decarburization (AOD) process, which gives very low carbon content, while nitrogen is often added to increase strength and improve pitting resistance of higher grades. The high-performance stainless steels, other than the extra-low interstitial or superferritic grades discussed above, fall into both duplex and austenitic categories.

Duplex Grades

The duplex grades characteristically contain molybdenum and have a structure of approximately 50% ferrite and 50% austenite because of the excess of ferrite-forming elements such as chromium and molybdenum. The duplex structure, combined with molybdenum, gives them improved resistance to chloride-induced corrosion (pitting, crevice corrosion, and SCC) in aqueous environments particularly.

It should be noted that the ferrite is not an unmixed blessing. Ferrite may be attacked selectively in reducing acids, sometimes aggravated by a galvanic influence of the austenite phase, while the sigma phase produced by thermal transformation (as by the heat of welding) is susceptible to attack by strongly oxidizing acids. The duplex structure is subject to 475°C (885°F) embrittlement, and has poor NDTT properties. Except for the temper embrittlement, these problems can be minimized through corrosion qualification tests and impact testing.

The prototype alloy was type 329 (S32900), followed by type 315 (S31500), both of which gave reasonably good service in many applications. However, these have been replaced with newer alloys, strengthened with nitrogen (which also replaces a small amount of nickel), and generally, are improved in fabrication and corrosion characteristics.

Today, the basic duplex steels are alloy 2205 (S31803) and alloy 255 (Ferralium† 255, S32550). Alloy 2205 has excellent stress corrosion cracking resistance, pitting, and crevice corrosion resistance exceeding type 317L in most environments, and more than twice the strength of type 316L. Alloy 2507 (S32750), with high nickel, molybdenum and nitrogen contents, has corrosion resistance similar to that of the 6% Mo austentitic stainless steels (see below), and with approximately three-times the strength of type 316L. The compositions of some widely used duplex grades are shown in Table 12.2.

The nitrogen-strengthened, high-chromium alloys find application in the more aggressive chloride environments (i.e., brackish and seawater).

Austenitic Grades

There are three major high-performance stainless steels which find application in the chemical process industries.

Alloy 22-13-5 (22%Cr-13%Ni-5 Mn-2%Mo-0.06 C, S20910) has chemical resistance equivalent to type 316L, better resistance to SCC, and much higher strength. It is useful for pump shafts, for example, to minimize corrosion fatigue in services where type 316L otherwise is used.

Alloy SX† (S32615) is nominally 18%Cr-18%Ni-1%Mo-2%Cu-5%Si and has outstanding resistance in concentrated sulfuric acid and some nitric acid applications, with its silicon content enhancing the surface film especially in strongly oxidizing acids.

Alloy 254 SMO† (S31254), referred to as a "superaustenitic" stainless steel, contains nominally 20%Cr-18%Ni-6%Mo-1%Cu-0.02%C. It is superior to the duplex grades in seawater and is used

Table 12.2

Duplex Stainless Steels, nominal wt. %

Name	UNS	Cr	Ni	Mo	Cu	C (max)	N	Other
Alloy 2304	S32304	23	4	—	—	0.02	0.10	—
Type 329	S32900	26	3.5	1.5	—	0.20	—	—
Alloy 3RE60	S31500	19	4.5	2.5	—	0.03	—	Si 1.5
Ferralium† 255(A)	S32550	26	6	3.0	1.5-2.5	0.04	0.15	—
Alloy 2205(B)	S31803	22	5.5	3.0	—	0.03	0.18	—
Zeron† 100(C)	S32760	25	7.5	3.8	0.7	0.03	0.25	W 0.75
Alloy 2507	S32750	25	7	4.0	—	0.03	0.25	—
Alloy 47N	S31260	25	6	3.0	< 0.5	0.03	0.20	—
Alloy 7-MoPLUS†	S32950	28	4.5	2.0	—	0.03	0.20	—

(A) Available in cast form.
(B) UNS J92205 in cast form.
(C) UNS J99380 in cast form.

in applications requiring both resistance to general corrosion and to chloride pitting and SCC. (*Note*: A similar family of 6%Mo superaustenitic alloys have different UNS designations—namely, alloys AL-6XN† is N08367 and alloys 1925hMo [also named alloy 926] and 25-6MO are N08926; see Chapter 13.)

12.3 Stainless Castings

It is important to understand that stainless steel castings may differ significantly from their wrought counterparts. One should never specify "cast type 316," for example, because a foundry might pour just that from bar-stock (to meet your specification) without regard to the proper balance of constituents. Table 12.3 lists the American Casting Institute (ACI) and UNS designations for a number of corrosion-resisting (C) and heat-resisting (H) castings. The number associated with the alphabetical identification of heat-resisting castings (e.g., HK-40 [J94204]) indicates the *mid-range* of the carbon content (± 0.05%). The 5% silicon steel SX† (S32615) is also available in cast form.

Table 12.3

Designations for Stainless Steel Castings, nominal wt. %

ACI	UNS	Cr	Ni (max)	Other
CA-*nn*(A)	–	12	1	C
CB-*nn*	–	19	2	C
CD-4MCu(B)	J93370	26	6	C 0.04 max., Mo 2, Cu 3
CF-8	J92600	19	11	C 0.08 max
CF-3	J92500	18	12	C 0.03 max
CF-8M	J92900	19	12	C 0.08 max, Mo 2
CF-3M	J92800	18	13	C 0.03 max, Mo 2
CH-20	J93402	23	15	C 0.20 max
CK-20	J94202	24	22	C 0.20 max
CN-7M	N08007	20	31	C 0.07 max., Mo 2, Cu 3
HK-*nn*(C)	–	24	20	C 0.60 max

(A) C indicates Corrosion-Resistant; nn is maximum carbon.
(B) Duplex alloy.
(c) H indicates Heat-Resistant; nn is *middle* ± 0.05% carbon.

Suggested Resource Information

ASM Handbook, Vol. 1, *Properties and Selection: Irons, Steels, and High-Performance Alloys* (Materials Park, OH: ASM International, 1990).

ASM Handbook, Vol. 13, *Corrosion* (Materials Park, OH: ASM International, 1987).

J. Frick, ed., *Woldman's Engineering Alloys,* 7th Ed. (Materials Park, OH: ASM International, 1990).

The Metals Black Book, Vol. 1, *Ferrous Metals* (Edmonton, Alberta, CAN: CASTI Publishing, 1992); available from NACE International.

B. J. Moniz, W. I. Pollock, eds., *Process Industries Corrosion–The Theory and Practice* (Houston, TX: NACE International, 1988), pp. 415 and 427.

J. G. Parr, A. Hanson, R. A. Lula, *Stainless Steel* (Materials Park, OH: ASM International, 1985).

D. Peckner, I. M. Bernstein, *Stainless Steels* (New York, NY: McGraw-Hill, 1977).

A. J. Sedriks, *Corrosion of Stainless Steels* (New York, NY: Wiley-Interscience, 1979).

13

High-Performance
Nickel-Rich Alloys

The basic alloy in the group of high-performance nickel rich alloys is alloy 800 (UNS N08800). This is a nickel-chromium-iron alloy, nominally 30%Ni-20%Cr-balance Fe. Strictly neither an iron-based nor a nickel-based alloy (because neither element is present at more than 50% concentration), the UNS "N" designation is arbitrary, as explained below.

Alloy 800 has much better resistance to corrosion, oxidation, and chloride-induced attack than does type 304L, for example. However, it is *not* better than the type 300 series in resisting caustic cracking. A variant for high-temperature services is alloy 800H (N08810). The place of alloy 800 in the market for chloride environments is being taken over almost completely by molybdenum-bearing grades.

This latter category embraces a number of alloys, somewhat similar to austenitic stainless steels, that contain chromium, nickel, and molybdenum as the main alloying elements. The original such alloy was a casting of approximately 20%Cr-29%Ni-3%Mo-2%Cu, developed primarily for intermediate concentrations of sulfuric acid. This material, as first developed, could not be obtained in wrought form. Eventually, the addition of small amounts of rare earth alloying elements resulted in a wrought material for sheet, plate, strip, pipe, and tubing. Somewhat later, in order to improve resistance to chloride-induced SCC, the nickel was increased and the alloy became known as alloy 20Cb-3[†] (N08020), sometimes referred to generically as "alloy 20."

Because this alloy could be classified as neither iron based nor nickel based, ASTM elected arbitrarily to assign it to the nonferrous category and assign a UNS number in the Nxxxxx listing. This policy continued and other related or somewhat similar alloys

[†] Trade name

were assigned to the nonferrous category and given UNS "N" designations, e.g., alloy 825 with 38%Ni-20%Cr-3%Mo-2%Cu (UNS N08825). These alloys characteristically have chromium, nickel, molybdenum (and sometimes copper) totalling in excess of 50%, and are stabilized with columbium (niobium) or titanium, or in more modern developments made with very low carbon (or with both low carbon and stabilized), to resist IGA. (*Note:* There is a movement afoot in ASTM to categorize an alloy by its predominant alloying element even if it is less than 50% concentration. This may result in some high-performance alloys [such as alloy 904L (N08904)] being included in the ASTM ferrous category with their UNS "N" designations retained.)

Many years ago, if type 316L or 317L was not suitable for an oxidizing chemical service, only the early or later versions of "alloy 20" (N08020) were available before having to use the nickel-based alloy C (N10002) or C-276 (N10276). (Nickel-based alloys are discussed in Chapter 14.) When the corrosion problems were related to chloride, especially as regards pitting and crevice corrosion rather than SCC, N08020 was not really superior to type 316L or analogous stainless steels.

Finally, a group of "4 Mo" alloys were developed with 4 to 5% molybdenum content (thereby improving resistance to localized corrosion) and lower nickel content (still yet sufficient to prevent SCC). The Fe-Ni-Cr-Mo alloy 700 (N08700) and the Fe-Ni-Cr-Mo-Cu alloy 904L (N08904) were among the prototypical compositions in this group. The 1 to 2% copper addition with molybdenum (e.g., in alloy 904L) is almost as effective as a somewhat higher molybdenum content in alloys without copper.

13.1 Superaustenitics

13.1.1 Iron-Nickel-Chromium-Molybdenum-Copper Alloys

More recently, because of some corrosion-related problems with the "4% Mo" grades in severe chloride services, a newer group of "6% Mo" superaustentics alloys with very-low carbon contents became available with improved chloride resistance to both pitting and SCC. The nitrogen additions result in high strengths. Because some alloys have UNS "S" designations and others "N" designations, it is probably best to consider groups of alloys for major applications.

Table 13.1

Fe-Ni-Cr-Mo-Cu Alloys, nominal wt. %

Alloy	UNS	Fe	Ni	Cr	Mo	Cu	C (max)	Other
20Cb-3[†]	N08020	40	33	20	2.5	3	0.07	Cb
28	N08028	35	31	27	3.5	1	0.03	—
904L	N08904	45	26	21	4.5	2	0.02	—
925	N08925	46	25	20	6	1	0.02	—

While alloy 20Cb-3[†] (N08020) was primarily intended for sulfuric acid, the Ni-Fe-Cr-Mo-Cu alloy 825 (N08825) is sometimes as good or better, depending upon specific contaminants in intermediate strengths of sulfuric acid (see Chapter 23). On the other hand, attempts to provide increased resistance to chloride pitting and crevice corrosion prompted the development of the 25%Ni-20%Cr + Mo alloys.

Table 13.1 (shown above) compares Fe-Ni-Cr-Mo-Cu alloys with alloy 20Cb-3[†], arranged in increasing molybdenum content. The lower carbon contents result from AOD processing and other improvements in melting practices.

13.1.2 Iron-Nickel-Chromium-Molybdenum Alloys

Table 13.2 lists "6% Mo" superaustenitic alloys. They comprise alloys 254 SMO[†], AL-6XN[†], 1925hMo (also called alloy 926), 25-6MO, and 31. The higher molybdenum level of alloy 31 gives excellent resistance to fretting and crevice corrosion in neutral

Table 13.2

Fe-Ni-Cr-Mo Alloys (6% Mo Superaustenitics), nom. wt %

Alloy	UNS	Fe	Ni	Cr	Mo	Cu	C (max)	Other
254 SMO[†]	S31254	54	18	20	6.1	0.7	0.020	N 0.20[(A)]
AL-6XN[†]	N08367	48	24	20	6	—	0.030	N 0.20[(A)]
1925hMo 926 25-6MO	N08926	46	25	20	6.2	1	0.030	N 0.20[(A)]
31	N08031	33	31	27	6.5	1	0.020	N 0.20[(A)]
654 SMO[†]	S32654	42	22	24	7.3	0.5	0.020	N 0.50[(A)]

[(A)] Nitrogen added for increased strength.

and aqueous acid solutions. Also included is alloy 654 SMO†, a new 7+% Mo alloy, with corrosion resistance associated with nickel-based alloys and with about double the strength of type 316L.

13.2 High-Nickel Austenitics

13.2.1 Nickel-Iron-Chromium-Molybdenum-Copper-Alloys

A number of alloys are apparently compromises, more costly than those alloys with iron as the predominant alloying element, intended to resist sulfuric acid and SCC, while simultaneously showing improved resistance to pitting and crevice corrosion.

Table 13.3 (shown below) gives compositions of two Ni-Fe-Cr-Mo-Cu alloys, with alloy 825 included for comparison.

13.2.2 Nickel-Chromium-Iron-Molybdenum-Copper Alloys

Among the high-performance alloys, one was developed years ago with the specific purpose of resisting the wet phosphoric acid process. The original alloy F evolved successively into columbium–containing alloy G (N06007), then to a low-carbon *and* columbium-stabilized variant alloy G-3 (N06985), and ultimately to the current chromium-enriched alloy G-30† (N06030). (See Table 13.4 and Chapter 14.2.3.)

13.3 Ranking Alloys

It must be emphasized, that one alloy cannot be ranked over another until the conditions and chemistry of exposure are defined. For example, type 316L is the *optimum* alloy in production of ethylene diamine by the ethylene dichloride route, since type 317L

Table 13.3
Ni-Fe-Cr-Mo-Cu Alloys, nominal wt. %

Alloy	UNS	Ni	Fe	Cr	Mo	Cu	C (max)	Other
825	N08825	42	30	22	3	2	0.05	Ti
20Mo-4†	N08024	38	33	23	4	1	0.03	Cb
20Mo-6†	N08026	35	31	24	6	3	0.03	—

Table 13.4
Ni-Cr-Fe-Mo-Cu Alloys, nominal wt. %

Alloy	UNS	Cr	Ni	Fe	Mo	Cu	C (max)	Other
G	N06007	22	45	20	6.5	2	0.05	Cb 2.0
G-3	N06985	22	48	18	7	2	0.015	Cb 0.3
G-30[†]	N06030	30	45	15	5.5	2	0.03	Cb 0.7, W 2.5

offers no better corrosion resistance and alloy 20Cb-3[†] shows less resistance by virtue of its high nickel content which can be complexed by hot amines. Furthermore, alloy G-30[†] resists nitric-hydrofluoric acid mixtures better than the Ni-Cr-Mo alloy C-276 (N10276). The relative resistance of alloy G-30[†] and alloy 28 (N08028) in wet process phosphoric acid varies with the chemical composition of the raw phosphate rock (especially as to contaminants) at various geographical locations.

On-going alloy developments can be expected and one must try to remain abreast through the available manufacturers' and technical literature.

Suggested Resource Information

ASM Handbook, Vol. 13, *Corrosion* (Materials Park, OH: ASM International 1987).

R. M. Davison, J. D. Redmond, "Practical Guide to Using 6 Mo Austenitic Stainless Steels," *Materials Performance* 27, 12 (December, 1988), p. 39.

The Metals Red Book, Vol. 2, *Nonferrous Metals* (Edmonton, Alberta, CAN: CASTI Publishing, 1993).

B. J. Moniz, W. I. Pollock, eds., *Process Industries Corrosion–The Theory and Practice* (Houston, TX: NACE International, 1988), p. 427.

14

Nickel and Its Alloys

The family of nickel alloys is among the most important because they resist corrosion in a wide variety of environments. The alloys can be divided into two groups: (1) those which depend primarily on the inherent characteristics of nickel itself (plus the influence of certain alloying additions); and (2) those which employ chromium as a major element to develop a passive film analogous to that which forms on iron-chromium alloys such as stainless steels.

14.1 Nickel Alloys

14.1.1 Nickel

Commercially pure nickel (e.g., alloy 200 [UNS N02200]), is a white, *magnetic* metal very similar to copper in its other physical and mechanical properties. The *Curie point*, i.e., the temperature at which it loses its magnetism, fluctuates with the nature and extent of alloy additions, rising with increased iron and cobalt concentrations, falling with addition of copper, silicon and most other elements. Nickel is also, of course, an important alloying element in other families of corrosion-resistant alloys (e.g., alloy cast irons, cupronickels, stainless steels).

Nickel is used alone as a material of construction, as a cladding on a steel substrate, and as a plating on steel or other less noble metals. The plating may be deposited either by electrochemical techniques or as *electroless* nickel plating (ENP) deposited by a chemical reduction process.

In addition to alloy 200, there are a number of alloy modifications developed for increased strength, hardness, resistance to galling, and improved corrosion resistance. The several variants of nickel are substantially equivalent in corrosion resistance, but the low-carbon alloy 201 (0.02% carbon maximum, N02201) is preferred for service above 300°C (570°F), as in caustic evaporators.

A major application of nickel is in the production of high-purity caustic in the 50 to 75% concentration range. The sodium hydroxide is produced in nickel or nickel-clad evaporators to meet the rigorous requirements of the rayon, soap, and other manufacturing industries for iron-free, copper-free caustic.

Nickel, like copper, is *not* resistant to ammonia or its derivatives in the simultaneous presence of air, oxygen, or other oxidants. This is due to the formation of soluble complex ions, similar to the copper-ammonium complex.

Nickel is moderately resistant to acids (since it does not liberate hydrogen), except under oxidizing conditions. Oxidizing cations (e.g., ferric or cupric ions) or anions (e.g., nitrates, nitrites) cause rapid attack.

Nickel-plated steel, particularly of the electroless variety (actually a nickel/nickel phosphide alloy), is used for the shipment and storage of many chemicals which require protection against iron contamination. Electroless nickel plating can be hardened by heat treatment for wear resistance approaching Cr-plate.

14.1.2 Alloy 400 (Monel[†])

Monel[†], or alloy 400 (UNS N04400), is a well-known and widely used alloy containing approximately 30% copper and up to 2.5% iron. Alloy 400 is readily forged, worked, cast, and welded (or brazed), and has approximately the same machinability as steel. Its Curie point is very close to room temperature, hence it can lose its magnetism when merely warmed by the sun, only to recover it upon slight cooling. (Cast Monel[†] has a higher Curie point because of increased silicon content.) There are a number of variants (e.g., precipitation-hardened N05500) for improved strength, hardness, machinability or resistance to galling.

Corrosion resistance is good in many (although not all) natural waters, including seawater. Both alloys 200 and 400 can be attacked in *soft* waters (see Chapter 19), having critical ratios of oxygen to carbon dioxide. Alloy 400, however, is the standard material of construction for watermeter parts, pumps, valves, strainers, etc. A substantial amount is used in fabrication of hot-

† Trade name

water tanks. It should be noted, however, that under hot-wall conditions (i.e., in coolers or condensers), some fresh waters will cause pitting, even though otherwise non-corrosive.

Alloy 400 and its variants will resist both acids and alkalis, in the absence of oxidizing agents. As one would expect, these nickel-copper alloys are subject to corrosion by ammonia and its derivatives in the presence of oxygen or oxidants.

These alloys are subject to environmental cracking under some conditions (e.g., SCC by hydrofluoric acid vapors plus air due to the influence of cupric fluoride, and by hot concentrated caustic; HAC by atomic hydrogen, in the cold-worked condition; LMC by mercury or its salts). For HF SCC of alloy 400 by cupric fluoride, thermal stress relief is only partially effective.

14.1.3 Nickel-Molybdenum Alloys

Alloy B (UNS N10001) is an alloy of nickel with approximately 28% molybdenum, developed to resist hydrochloric acid up to the atmospheric boiling point. This resistance is, however, severely impaired by the presence of oxidizing contaminants (e.g., ferric ions in muriatic acid). As one would expect, the alloy resists reducing acids quite well, and oxidizing acids not at all. Alkali resistance is also good, but use of such an expensive alloy is not warranted. Like other high-nickel alloys, it is susceptible to attack by amines, in the presence of air or oxidants, due to the nickel-ammonium type complex.

Because of susceptibility to IGA in the HAZ after welding in some environments, a low-carbon variant has been developed. Alloy B-2 (UNS N10665) is replacing alloy B in most applications. An improved version, alloy B-3[†] (UNS N10675), has been introduced. It is specially formulated to achieve a level of thermal stability superior to alloy B-2. Improved thermal stability provides superior fabricability, e.g., ease of heat treatment procedures. Alloy B-4[†] (N10629), also recently introduced, has high resistance to SCC in the aged condition and no embrittlement during thermal-mechanical processing.

14.2 Chromium-Bearing Alloys

14.2.1 Nickel-Chromium Alloys

The prototype alloy in this category is alloy 600 (N06600), often called just Inconel[†] with no designation following the trade name, which contains approximately 76% nickel, 16% chromium, and 8% iron. The presence of chromium gives it a passive film and corrosion characteristics analogous to those of stainless steels. A variant is alloy 601 (N06601), with a somewhat lower carbon content and a somewhat higher chromium content.

Alloy 600 is used as a heating coil in caustic evaporators, unless traces of Cr(VI) ions are objectionable in the product. Nevertheless, at high stress levels, it is subject to SCC in strong caustic on prolonged exposure above approximately 300°C (570°F).

Alloy 600 can replace austenitic stainless steel in some chloride environments, but it can be pitted (although not cracked) under severe conditions. It is at least an order of magnitude better than type 300 series stainless steels in steam contaminated with *both* chlorides and caustic, and has been widely used to replace type 321 (S32100) stainless steel for corrugated expansion joints in such service.

Alloy 600 has excellent resistance to halogens and halogen acid anhydrides at elevated temperatures. However, as with all nickel alloys, it is subject to LMC by sulfur contamination at elevated temperatures (e.g., in welding, where the materials must be scrupulously clean and free of any sulfur-bearing contaminating species). A molybdenum-bearing variant is discussed below.

14.2.2 Nickel-Chromium-Molybdenum Alloys

The prototype alloy in this category was alloy C (N10002) containing approximately 15% each of chromium and molybdenum to enhance acid resistance under oxidizing conditions. An improved alloy of greater resistance to IGA is alloy C-276 (N10276), often called just Hastelloy[†] with no designation following the trade name. Alloy C-276 has virtually replaced alloy C. A more stable variant is alloy C-4 (N06455). (When alloy 600 is improved by increasing the chromium content to approximately 21% plus 8% molybdenum, alloy 625 [N06625] is obtained.) Alloy development continues in this area, namely, alloys C-22[†] (N06022) and 59 (N06059). Table 14.1 gives compositions of Ni-Cr-Mo alloys.

Table 14.1

Nickel-Chromium-Molybdenum Alloys, nominal wt. %

Alloy	UNS	Ni	Cr	Mo	W	Fe	C (max)	Other
C-276	N10276	57	16	16	4	5	0.01	—
C-4	N06455	65	16	16	—	2	0.01	—
C-22†	N06022	56	22	13	3	3	0.01	—
59	N06059	59	23	16	—	1	0.01	—
625	N06625	61	21	8	—	4	0.10	Cb 3.5

These alloys are outstanding in resistance to hot seawater, aggressive organic acids (e.g., boiling formic or acetic acid mixtures) and hot oxidizing acids (e.g., mixtures of nitric with hydrochloric). Alloy 625 is usually, although not always, somewhat less resistant than the 15-16% molybdenum grades in chemical services. Ni-Cr-Mo alloys are also sometimes employed for severe SCC environments (e.g., centrifuges for high-chloride resin or polymer mixtures, such as polyvinyl chloride [PVC]) where the super-stainless steels or high-performance nickel-rich alloys might be suspect or of too low a mechanical strength.

Again, as in Chapter 13.3, it should be noted that superiority cannot be claimed for one Ni-Cr-Mo alloy over another (nor even over alloys of lesser content of elements conveying or enhancing corrosion resistance) without reference to specific chemical environments.

Suggested Resource Information

ASM Handbook, Vol. 13, *Corrosion* (Materials Park, OH: ASM International, 1987).

W. Z. Friend, *Corrosion of Nickel and Nickel-Base Alloys* (New York, NY: Wiley-Interscience, 1980).

The Metals Red Book, Vol. 2, *Nonferrous Metals* (Edmonton, Alberta, CAN: CASTI Publishing, 1993).

B. J. Moniz, W. I. Pollock, eds., *Process Industries Corrosion–The Theory and Practice* (Houston, TX: NACE International, 1988), p. 461.

15

Lead, Tin and Zinc

These three metals are grouped together for discussion because they share an amphoteric nature (i.e., they can be attacked by both acids and alkalis), because they are all employed as metallic coatings, and because it is convenient to separate them from other nonferrous metals, such as copper and nickel, more commonly employed in corrosive service.

15.1 Lead and Its Alloys

Lead is a weak, heavy metal (almost half again as heavy as steel) which has difficulty supporting even its own weight. However, it has enjoyed widespread industrial use, particularly because of its resistance to sulfuric acid. In nature, it is usually associated with copper and silver. Applications of lead have been greatly diminished in modern practice, because of toxicity problems associated with its joining ("lead-burning").

15.1.1 Types of Lead

There are a number of lead alloys developed specifically to increase endurance limit and hardness (or ability to work-harden) and to prevent excessive grain growth.

Chemical lead is lead with traces of copper and silver left in it. It is not economical to recover the silver, and the copper content is thought to improve general corrosion resistance.

Antimonial lead (also called *hard lead*) has been alloyed with from two to six percent antimony to improve the mechanical properties. This is effective in services up to approximately 93°C (200°F), above which both strength and corrosion resistance rapidly diminish.

Tellurium lead has been strengthened by the addition of a fraction of a percent of tellurium. It will work-harden under strain, thus it has better resistance to fatigue failure induced by vibration.

There are also proprietary alloys (e.g., Asarcon[†]) to which copper and other elements have been deliberately added for improved corrosion and creep resistance. Nalco[†] has been alloyed specifically for improved corrosion resistance in chromic acid plating baths.

15.1.2 Forms

Lead is available in pipe, tubing, and valves, as well as sheet for process vessels. Sheet lead is also used as a membrane in acid brick-lined vessels. Lead-clad steel plate is also available. Because it will not easily support its own weight, lead is subject to creep and *static fatigue*. Creep is characterized by stretching under constant load. Static fatigue is characterized by a "crazing" effect and, sometimes, by IGA.

For these reasons, lead vessels are usually externally supported by wooden or steel structures, or fabricated from clad metal (i.e., homogeneously bonded lead-clad steel). Since the steel supplies the mechanical strength, any grade of lead may be used as the cladding, at the discretion of the manufacturer.

15.1.3 Corrosion Resistance

Lead is an amphoteric metal, i.e. susceptible to attack by both acids and alkalis under certain conditions. That it has excellent resistance in many environments is due largely to the insolubility of some of its corrosion products, which tend to form protective films in the appropriate services.

Lead tends to resist sulfuric, sulfurous, chromic, and phosphoric acids, and *cold* hydrofluoric acid (see Chapters 19 to 32 on specific environments). It is attacked by hydrochloric acid and by nitric acid, as well as organic acids, if they are dilute or if they contain oxidizing agents.

The usefulness of lead in caustic is limited to concentrations of not more than 10% up to approximately 90°C (195°F). It suffers little attack in cold, strong amines, but is attacked by dilute aqueous amine solutions.

[†] Trade name

Lead will resist most natural waters, but is not suitable in soft aggressive waters. It must not be used to handle potable (i.e., drinking) water, because of the toxicity of lead salts.

Although lead is anodic initially to more highly alloyed materials, it may become cathodic in time due to the film of insoluble corrosion products formed on its surface. As an example, alloys 20-Cb3[†] (UNS N08020) and C-276 (N10276) valves may suffer accelerated attack in lead piping systems in sulfuric acid services unless electrically isolated from the piping.

15.2 Tin

Tin is of somewhat limited availability in the United States, since all of it must be imported. It is a major alloying element in "phosphor" and other types of bronzes, but its major usage as a material of construction is as tin-plate and solder.

The use of tin as a packaging medium (e.g., as tooth-paste tubes, foil, etc.) has largely given way before other materials, such as plastics. However, tin-plate (e.g., over steel) is produced both as a hot-dipped and as an electroplated coating. Attempts to replace "tin" cans for canned peaches and apricots have been unsuccessful from a marketing standpoint, because the organotin complex (which gives the products their characteristic taste) does not form in plastic or glass containers.

Tin is encountered to a limited extent in *terneplate* (a hot-dipped steel for roofing purposes). The most common solder is a 50-50 lead alloy.

15.3 Zinc

Zinc is a major alloying element in brass. For materials of construction, the major engineering useage is as the hot-dipped galvanized steel and as a sacrificial pigment in certain types of paints. Electroplated zinc coatings are also encountered. The electrochemical protection is directly proportional to the thickness of the zinc, so the method of application to steel is selected on the basis of feasibility and economics.

Zinc die-castings are used for some artifacts not intended for exposure to corrosion. Zinc anodes are widely used for CP in saltwater services, such as to protect rudder irons from corrosion and propellers from cavitation.

Suggested Resource Information

ASM Handbook, Vol. 2, *Properties and Selection–Nonferrous Alloys and Special Purpose Materials* (Materials Park, OH: ASM International, 1991).

The Metals Red Book, Vol. 2, *Nonferrous Metals* (Edmonton, Alberta, CAN: CASTI Publishing, 1993).

16

Copper and Its Alloys

Copper and its alloys have been known and utilized for more than 6,000 years. The Bronze Age antedates the Iron Age by approximately 1,500 years. The corrosion resistance of copper alloys was recognized almost from the beginning, as evidenced by their use in containers for food and drink and for long-lasting statuary. Apparently, initial workings of hammered copper were followed by crude castings of lean bronze from impure copper ores. Finally, true bronzes, e.g., tin (phosphor) bronzes, appeared. In modern times, highly purified coppers, free of all but traces of other elements, as well as copper alloys compounded for specific properties, are widely used in industry. Table 16.1 (on following page) lists the nomenclature and approximate compositions of some of the more important wrought and cast copper alloys.

16.1 Copper

There are basically two types of copper of interest in corrosion applications. Electrolytic tough pitch copper (ETP; C11000) is used in sheet metal work and process equipment. Phosphorus-deoxidized coppers, e.g., high residual phosphorus (DHP) UNS C12200 and low residual phosphorus (DLP) C12000, have excellent response to hot-drawing (e.g., as for heat exchanger tubing), flaring, flanging, spinning, and welding. (Oxygen-free electronic [OFE] or high-conductivity copper [C10100] is used only for electrical devices.)

The great majority of copper alloys cannot be hardened by heat treatment. They are hardened by cold-work, such as hammering, as in primitive times. (There is no "lost secret of the ancients" for hardening copper.) With subsequent reheating, there are several "tempers" available (e.g., quarter-hard, half-hard) with different capabilities for flaring or rolling. Modern technology has given us a few precipitation-hardening alloys (copper plus chromium or zirconium) and a grade which is dispersion-strengthened copper (e.g., C15710).

The corrosion resistance and mechanical properties of the several types of copper (in the unhardened condition) are substantially the same. Copper will resist most natural waters within certain velocity limitations (except soft, aggressive waters), as well as both acids and alkalis (in the absence of dissolved oxygen or other oxidizing species). It should be noted that the ionic corrosion product (specifically the *cupric* ion) is itself an oxidant, so that acid solutions become more corrosive to copper and its alloys as corrosion products accumulate. As is evident from the electromotive series, copper will not displace hydrogen from acids. It is attacked only when the cathodic reaction is the reduction of dissolved oxygen, of metallic cations or of the anion of the acid, as in oxidizing acids (e.g., nitric acid). The copper alloys are corroded by ammonia and amines, in the presence of oxygen or oxidizing agents, because of the soluble copper-ammonium or copper-amine complexes.

Table 16.1

Copper Alloys, nominal wt. %

Name	UNS	Cu	Zn	Sn	Al	Ni	Other
Wrought							
Copper	C11000	99.9	—	—	—	—	—
Red Brass	C23000	85	15	—	—	—	—
Yellow Brass	C27000	69	31	—	—	—	
Admiralty B	C44300	72	27	1	—	—	As 0.1
Phosphor Bronze	C52400	90	—	10	—	—	P 0.3
Al Bronze D	C61400	90	0.2	—	7	—	Fe 3
High Si Bronze	C65500	95	1.5	—	—	0.6	Si 3
90-10 Cu-Ni	C70600	86	1.0	—	—	10	Fe 1.5
70-30 Cu-Ni	C71500	68	1.0	—	—	30	Fe 1
85-15 Cu-Ni	C72200	83	1.0	—	—	15	Cr, Fe
Castings							
Ounce Metal	C83600	85	5	5	—	—	Pb 5
Mn Bronze	C86500	57	40	1	—	1	Mn 1
G Bronze	C90500	87	2	10	—	1	—

16.2 Brass

The term *brass* refers specifically to copper alloys in which the major alloying element is zinc. This is an exception to the use of the term *bronze* (see Chapter 16.3), which includes almost any alloy of copper with a significant amount of alloying element (e.g., tin, silicon, aluminum, nickel).

Modern brasses contain anywhere from 5 to 45% zinc. As the zinc content is increased, the characteristic red brass color is first observed, up to 10 to 15% zinc (C23000). The yellow brass next appears commercially in the range 25 to 30% zinc (C26000). As the zinc concentration is further increased to approximately 38% (C46400, naval brass), the red color returns (due to the formation of a secondary beta phase in lieu of alpha brass) until, at approximately 42% zinc, the color is again a close match to 90-10 red brass. Small amounts of lead, which is insoluble in copper, may be added to brasses or bronzes to improve machinability and lubricity (as in bearing alloys). Unfortunately, the lead particles can also function as corrosion initiation sites in some environments.

A potential corrosion problem with brasses containing more than approximately 15% zinc is *dezincification*, a form of parting corrosion, in some environments. This is a problem in many waters, ranging from potable water to seawater. To combat this type of attack in seawater, the British developed admiralty metal by adding 1% tin to a 70-30 yellow brass. In modern practice, the beneficial effect of tin is increased synergistically by additions of approximately 0.2% arsenic, antimony, or phosphorus. These additions provide the three (roughly equivalent) alloys; admiralty B (C44300), C (C44000), and D (C44500), respectively. Inhibiting additions of tin and arsenic are also made to other high-zinc alloys (e.g., aluminum brass, C68700; leaded muntz metal, C36600) intended for service in corrosive waters or other aggressive environments.

Another major weakness of brass and of other high strength copper alloys is a susceptibility to environment cracking, especially SCC and LMC (e.g., by mercury), as described in Chapter 8.

16.3 Bronze

There are four general categories of bronzes which are of interest from the corrosion engineering standpoint, although there are a great number of variants used in materials engineering of metal products. Bronzes and brasses (the nomenclature has always been rather loose until the advent of the UNS) have been widely used for naval and military applications, as well as architectural and machinery uses. Beryllium bronze is employed for improved hardness and wear-resistance (e.g., as in nonsparking tools).

16.3.1 Phosphor Bronze

These tin bronzes have taken their popular name from the use of phosphorus as a deoxidizer. The tin, usually used in concentrations of up to approximately 10% (e.g., C52400), confers strength and increases corrosion resistance in some environments. Casting bronzes may be more complex, e.g., leaded phosphor bronze (ounce metal, C54400 [85%Cu-5%Sn-5%Zn-5%Pb]), because of the variety of properties desired, such as fluidity in casting, strength, galling resistance, machinability, and corrosion resistance.

16.3.2 Silicon Bronze

Silicon bronzes are copper plus 1 to 3% silicon (e.g., C64900, C65500), noted particularly for strength and cryogenic suitability. However, these alloys are hot-short, because of the presence of a low-melting constituent at the grain boundaries. At 700 to 800°C (1,290 to 1,475°F), as from welding or hot-work, microfissuring occurs under stress. After welding, brazing or hot-working, the microfissures may become apparent as macrocracks after only a few hours. On the other hand, they may take several years to become apparent. The intergranular cracking is difficult to distinguish from SCC under the optical microscope. (The alloys are susceptible to SCC in ammonia or even uncontaminated steam.) LMC, as by mercury or its salts, is also difficult to distinguish from hot-short cracking, although the scanning electron microscope and microprobe analysis techniques should enable one to do so.

16.3.3 Aluminum Bronze

Aluminum bronzes are alloys of copper and aluminum in the

range from 5 to 10 percent (e.g., C61000). These are good, corrosion-resistant alloys, with remarkable strength at moderately elevated temperatures, compared with other copper alloys. They are rated for service to approximately 260°C (500°F) by the ASME Boiler and Pressure Vessel Code.[1] A proprietary grade, C61900 or Ampco 8[†] has been developed, which is especially inhibited for resistance to SCC by steam or ammonia.

16.3.4 Cupronickels

The cupronickels, which are not commonly referred to as bronzes, are a series of alloys ranging from 90-10 Cu-Ni (C70600) and 80-20 Cu-Ni (C71000) to 70-30 Cu-Ni (C71500). These have superior resistance to seawater (e.g., as condenser tubes), and the resistance can be improved by the addition of small amounts of iron (1.0 to 1.8%). There is a relatively new alloy competitive with C70600 in seawater, an iron- and chromium-modified 85-15 cupronickel (C72200), which is reported to be superior as regards corrosion, heat-transfer and biofouling. Cupronickels also have relatively good high temperature properties, and better resistance to erosion and cavitation than other copper alloys.

Alloy 400 and its variants will resist both acids and alkalis, in the absence of oxidizing agents. As one would expect, these nickel-copper alloys are subject to corrosion by ammonia and its derivatives in the presence of oxygen or oxidants.

Reference

1. *ASME Boiler and Pressure Vessel Code* (New York, NY: ASME).

Suggested Resource Information

ASM Handbook, Vol. 2, *Properties and Selection: Nonferrous Alloys and Special-Purpose Materials* (Materials Park, OH: ASM International, 1991).

The Metals Red Book, Vol. 2, *Nonferrous Metals* (Edmonton, Alberta, CAN: CASTI Publishing, 1993).

B. J. Moniz, W. I. Pollock, eds., *Process Industries Corrosion–The Theory and Practice* (Houston, TX: NACE International, 1988), p. 479.

[†] Trade name

17

Reactive, Refractory and Noble (Precious) Metals

Reactive metals are so named because of their affinity for hydrogen and oxygen, particularly at elevated temperatures (e.g., during welding). This is a misnomer to the extent that these metals and their alloys are corrosion-resistant in many severe environments. A refractory metal has a very high melting point. The *noble metals* (as opposed to *base metals*) are those which are found free and unreacted in the natural environment. These terms are somewhat obsolete; the noble metals are now generally called *precious metals*, because of their monetary value.

17.1 Reactive and Refractory Metals

The reactive metals of practical interest are titanium, tantalum and, zirconium. Tantalum is also classified as a *refractory metal*, because of its high melting range and high-temperature strength. All three are refractory in the sense of being difficult to win from their corresponding ores.

17.1.1 Titanium

Titanium is the ninth most abundant element on earth, but so reactive with the ordinary constituents of air at elevated temperatures that only in the 1950s did it become commercially available as an engineering material. As we learned to extract, refine, alloy and fabricate it, titanium became increasingly important. With a melting range of approximately 1,670°C (3,040°F), a low density (sp gr 4.5; approximately 60% that of steel), high strength (414 MPa [60,000 psi] minimum tensile strength) and good corrosion resistance, it showed great promise.

As discussed further below, it does have good corrosion resistance in strongly oxidizing environments, but *not* with reducing acids. Furthermore, the mechanical strength drops off rapidly with increasing temperature and its reactivity with hydrogen, oxygen, water vapor, and nitrogen effectively limits applications to less than approximately 535°C (995°F), except perhaps in controlled inert atmospheres. All welding, for example, must be done with GMAW, GTAW, or similar inert-gas processes. Because of the absolute exclusion of atmospheric contaminants required, field fabrication is somewhat difficult.

The strength of titanium can be increased by alloying (some alloys reaching 1,300 MPa or 190,000 psi), although at some cost in corrosion resistance. Titanium has good impact strength at cryogenic temperatures (although it must not be used in liquid oxygen because of the potential for explosion). It can be readily shaped and formed and is available in conventional forms.

The more important commercial grades of titanium and its alloys are listed in Table 17.1. Of these, Grade 2 (R50400) is the most commonly used, although Grade 1 (R50250) has optimum ductility and Grade 3 (R50550) higher strength.

The feature that moves titanium from its anodic position in the EMF series to a noble position in the Galvanic Series is the development of an inert, passive surface film in oxidizing environments. It is widely used in such strongly oxidizing media as wet chlorine and strong nitric acid. Titanium shows excellent resistance to seawater and other saline environments.

Titanium is generally not useful in non-oxidizing or reducing acids except in very dilute solutions (e.g., a few hundredths of a percent) or unless their redox potential is shifted by oxidizing contaminants (e.g., ferric ions, cupric ions, nitrites, etc.). Titanium is treacherous in hot organic acids (e.g., formic or acetic), in which it may suddenly lose its passivity and corrode at greatly accelerated rates. The addition of a small amount of platinum (e.g., grades 7, 11, and 16) greatly extends the tolerance for higher concentrations of reducing acids. In grade 12, small alloying additions of Ni and Mo improve resistance to crevice corrosion in brines and provides higher strength than unalloyed grades. Even moderately low corrosion rates can result in hydriding, with attendant loss of ductility, even before significant metal loss occurs. Titanium absolutely will not tolerate even trace amounts of fluorides, which cause severe hydriding and embrittlement as well as aggravated corrosion.

Table 17.1

Titanium Used in CPI[1]

ASTM Grade	UNS	Composition	Characteristics
1 (C.P.)[A]	R50250	Unalloyed Ti	Ductility, lower strength
2 (C.P.)[A]	R50400	Unalloyed Ti	Good balance of moderate strength and ductility, common workhorse alloy
3 (C.P.)[A]	R50550	Unalloyed Ti	Moderate strength
7 & 11	R52400 & R52250	Ti-0.15%Pd	Improved resistance to reducing acids and superior crevice corrosion resistance
16	—	Ti-0.05%Pd	Resistance similar to Grade 7, but at lower cost
12	R53400	Ti-0.3%Mo-0.8%Ni	Reasonable strength and improved crevice corrosion resistance
9	R56320	Ti-3%Al-2.5%V	Medium strength and superior pressure code design allowables
18	—	Ti-3%Al-2.5%V-0.05%Pd	Same as Grade 9, but with improved resistance to reducing acids and crevice corrosion
5	R56400	Ti-6%Al-4%V	High strength and toughness

[1] R. W. Schutz, *Materials Performance* 31, 10 (1992), p. 58.
[A] Chemically pure.

SCC has been observed in low molecular weight alcohols (e.g., methanol, ethanol), which tendency is inhibited by traces of water but aggravated by traces of halides or halogen acids. SCC in chlorinated solvents can be inhibited with small amounts of butylene and triethylamine. Titanium can also suffer cracking in strong nitric acid and in *dry* sodium chloride above approximately 300°C (570°F). HAC can occur in organic acids, due to galvanic effects of surface iron contamination unless the product is adequately pickled or anodized.

There is also a problem in strong oxidants under some conditions. Although titanium is excellent in wet chlorine, it can *burn* in dry chlorine (Chapter 26) unless a minimum water content of 2,000 ppm is maintained. Crevice corrosion may be a problem even in wet chlorine, because of depletion of oxidizing species within the crevice. It can also react catastrophically with *red fuming nitric acid* (RFNA) as discussed in Chapter 22. It can be *detonated* in liquid oxygen, in which its good low-temperature properties and low NDTT would otherwise make it attractive.

In short, titanium is a marvelously useful material of construction with a number of severe limitations. It is also relatively expensive, except in the form of thin-walled heat exchanger tubing which is produced in large quantities. Currently, it is competitive with nickel-based alloys.

17.1.2 Zirconium

Zirconium is between titanium and steel in density (sp gr. 6.5) and has a strength of approximately 345 MPa (50 ksi) in the annealed condition. The strength can be increased to approximately 550 MPa (80 ksi) by cold-work. The melting point is approximately 1,850°C (3,360°F).

Unless specifically removed, approximately 2% hafnium is present in zirconium (e.g., unalloyed zirconium R60701 and R60702). Although this does not impair corrosion resistance, it is objectionable in nuclear applications, for which an unalloyed reactor grade (R60001) is available. Other alloys, also reactor grade, are available (e.g., R60802, R60804, R60901). Zirconium is deliberately alloyed with aluminum, copper, molybdenum, niobium, tantalum, titanium, and tin to increase strength and corrosion resistance.

Zirconium and its alloys are easily fabricated, except for a hot-gas pick-up analogous to that to which titanium is susceptible. Thoroughly shielded inert gas welding is required. Even nitrogen pick-up of as little as 500 ppm, while not affecting mechanical properties, will diminish corrosion resistance in supercritical waters in nuclear reactors. The materials are also notch-sensitive, and threaded parts such as tray support rods in distillation columns are susceptible to breakage if the column should be "bumped" in service.

Zirconium is quite acid-resistant, withstanding not only oxidizing acids like nitric acid but also up to 60% phosphoric and up to 70% sulfuric acid, depending upon contaminants. It will also resist hydrochloric acid in the *liquid* phase, although not the hot vapors. In immersion service, arc-melted zirconium will resist HCl of all concentrations up to the atmospheric boiling point. However, if the material is not of high enough purity, corrosion of the weld HAZ may be a problem. This situation is exacerbated by the presence of oxidizing contaminants (e.g., ferric or cupric ions, nitrates, hypochlorites, chlorine, etc.). Zirconium is *attacked* by hydrofluoric and hydrobromic acids, but highly resistant to hot organic acids (e.g., acetic acid). In reducing acids contaminated with oxidizing ions (particularly hydrochloric and sulfuric acid solutions), *pyrophoric* corrosion products can be formed, which are a potential ignition source if flammable vapors are about (e.g., hydrogen emitted from pickling baths using zirconium heating coils).

Zirconium is caustic-resistant, but its use is rarely justified in such service. Hot molten metals (e.g., mercury, lead, bismuth) attack zirconium and its alloys.

Zirconium is more expensive than titanium, even in heat-exchanger tubing, but is competitive with titanium and the nickel-based alloys in many instances. It finds application in intermediate sulfuric acid strength services and in nuclear energy systems.

17.1.3 Tantalum

Tantalum is a very heavy, refractory metal, with a density more than twice that of steel (sp. gr. 16.6). Its melting point is almost 3,000°C (5,430°F). The tensile strength is approximately 345 MPa (50,000 psi), which can be approximately doubled by cold-work. Like titanium and zirconium, it is easy to fabricate, contingent upon completely inert conditions during welding, but is even more susceptible to hydrogen pick-up. Even in a simple galvanic couple at room temperature (e.g., with carbon steel in an aqueous environment), it will be severely embrittled by absorption of over 700 times its own volume of nascent hydrogen.

The corrosion resistance of tantalum is very similar to that of glass, resisting most acids but being attacked by hydrofluoric acid and by caustic. Unlike glass, however, it is attacked by fuming sulfuric acid, chlorosulfonic acid, and by sulfur dioxide.

Because of its very high cost, tantalum is usually used only in very thin sections, as a lining or thin cladding, or as a uniquely pore-free electroplated coating on a copper or steel substrate such as for orifice plates). Tantalum-plated copper can be immersed in nitric acid, or tantalum-plated steel in hot concentrated sulfuric acid, with no attack of the substrate (because of the absence of holidays in the plating).

Orifice plates, bayonet heaters, heat exchangers, valves and tantalum-plated steel tubes are of interest in the process industries. Tantalum patches are used to repair holidays in glass-lined steel vessels, but *must* be electrically isolated from other metallic components in the vessel (to prevent embrittlement by hydrogen absorption as the cathode in a galvanic couple).

17.2 Noble (Precious) Metals

The noble (or precious) metals are very expensive. Nevertheless, they do find some application in process equipment, under special conditions. If fully resistant, their salvage value may easily offset the initial cost. If consumed in use, they may nevertheless pay for themselves on a cost-to-life basis, if they are truly needed for the service.

17.2.1 Silver

Silver was used for certain specific services before the development of modern nickel-based alloys and plastics. In everyday use (e.g., jewelry, silverware), the most common observation concerning silver is that it tends to tarnish. This brownish-black film is caused by traces of hydrogen sulfide in the atmosphere, but the film is fairly protective.

Silver is rapidly attacked by cyanides, and by oxidizing acids (e.g., nitric acid, strong sulfuric acid). It can form explosive compounds, called *azides*, with ammonia and its derivatives (whence the hazard from old ammoniacal silver nitrate solutions in the laboratory), but has been used successfully in some urea plants.

Silver resists organic acids, and, prior to the development of the nickel-chromium-molybdenum alloys (e.g., N10276), was used rather widely in such applications. Silver heating coils were used, for example, in type 316L (S31603) acetic acid storage tanks, where stainless coils give excessive iron contamination. Silver has also been used in handling and storage of phenols, acidic food-stuffs, and pharmaceuticals. It has been used for dilute hydrochloric acid, although ingress of air impairs its service life. Good resistance to hydrofluoric acid has led to its use in some applications related to the manufacture of fluorinated solvents (e.g., refrigerators and propellants).

Silver is usually resistant to high temperature caustics and alkalis, and has been used to handle molten alkaline salts.

17.2.2 Gold

Gold has been used for gaskets in certain severe services and, as the alloy of gold and platinum, for "spinnerettes" in the rayon industry. Gold is readily attacked by cyanides and by halogens. This is why, although it readily withstands nitric and hydrochloric acids separately, it is attacked by aqua regia (a 3:1 mixture of hydrochloric and nitric acids, which releases nascent chlorine). It does, however, resist sulfuric acid.

Gold will resist alkalis well, although the cost is seldom justified, but is attacked by alkaline sulfides in a manner analogous to silver.

17.2.3 Platinum

Although platinum is even more expensive than gold, it is used for electrical contacts in oxidizing atmospheres. It is very resistant to acids and alkalis except for aqua regia, halogens in general (i.e., chlorine, bromine, iodine) and halogen acids above 100°C (212°F).

All of the noble metals can be used as electroplated coatings but this is quite dangerous for corrosive services because they are strongly cathodic to any substrate of baser metals, leading to localized corrosion at holidays in the plating.

Suggested Resource Information

ASM Handbook, Vol. 2, *Properties and Selection: Nonferrous Alloys and Special–Purpose Materials* (Materials Park, OH: ASM International, 1991).

ASM Handbook, Vol. 13, *Corrosion* (Materials Park, OH: ASM International, 1987).

D. R. McIntyre, C. P. Dillon, *Guidelines for Preventing Stress Corrosion Cracking in the Chemical Process Industries.* MTI Publication No. 15 (Houston, TX: NACE International, 1985).

D. R. McIntyre, C. P. Dillon, *Pyrophoric Behavior and Combustion of Reactive Metal,* MTI Publication No. 32 (Houston, TX: NACE International, 1988).

The Metals Red Book, Vol. 2, *Nonferrous Metals* (Edmonton, Alberta, CAN: CASTI Publishing, 1993).

B. J. Moniz, W. I. Pollock, eds., *Process Industries Corrosion–The Theory and Practice* (Houston, TX: NACE International, 1988), pp. 503 and 545.

R. W. Schutz, "Understanding and Preventing Crevice Corrosion of Titanium Alloys," Part I, *Materials Performance* 31, 10 (October, 1992), p. 57; Part II, *Materials Performance* 31, 11 (November, 1992), p. 54.

18

Nonmetallic Materials

There are a variety of organic and inorganic materials, in addition to metals and alloys, with which the working engineer should have at least a nodding acquaintance. It is important to recognize that, although the nonmetallics do not corrode in the sense of metals and alloys, they are not thereby cure-alls, because they have their own problems as to chemical resistance, as well as certain strength and temperature limitations. Paints and coatings are materials, but not of construction, and are considered separately in Chapter 36.

18.1 Plastics

Since the development of celluloid in 1869, engineers have been interested in potential industrial applications of plastic materials. There are hundreds of plastics available today, their use (or potential use) determined by particular chemical characteristics and physical and mechanical properties.

A *plastic* is a material that contains as an essential ingredient an organic compound of high molecular weight, is a solid in the finished state, and, at some stage in its manufacture or processing, can be shaped or formed. In common parlance, a plastic is a *thermoplastic* if it can be softened and reshaped, without damage or degradation, under the influence of heat. If not, it is called a *thermosetting* material. However, most people refer rather loosely to both categories of resins as "plastics."

18.1.1 Thermoplastics

The thermoplastics are the most widely used organic materials of construction. When not modified or reinforced, their mechanical properties are the lowest of the several materials of interest, and rapid diminution of strength is encountered with only moderate temperature increases. However, the physical and mechanical properties are entirely adequate for many applications.

Properties

(1) *Strength.* The tensile and flexural strengths of the thermoplastic materials are rather poor, dropping off rapidly with increasing temperature. However, the use of organic or inorganic reinforcing agents can upgrade the strength, as can "alloying" with other resinous materials.

(2) *Hardness.* The original hardness of plastics is relatively unimportant, because resistance to wear and abrasion is more related to resilience than hardness. However, *changes* in hardness are an important criteria in evaluating chemical resistance. Changes in hardness, plus or minus, after exposure indicate a significant chemical attack (as do changes in volume, indicative of absorption or dissolution).

(3) *Creep.* Distension under constant strain at ambient temperature is roughly comparable to that of carbon steel at approximately 500°C (930°F). Consideration must be given to support of thermoplastic structures, particularly if exposed to heat, even direct sunlight.

(4) *Modulus.* This property, as well as thermal expansion (see below), deserves the greatest consideration in design of thermoplastic structures. The modulus is low and, in flexure, there really is no straight-line relationship between stress and strain.

(5) *Thermal Expansion.* Although the data are readily available in the literature, all too often the designer fails to provide adequate allowance for thermal expansion (which may be fifteen times that of an austenitic stainless steel). Reinforced materials like fiberglass-reinforced plastics (FRP) have much lower coefficients than do the thermoplastics. There

must be adequate room for the structure to expand and contract, while still providing adequate mechanical support.

(6) *Impact.* Thermoplastic materials, as a group, are tough materials. However, most of the materials are notch-sensitive, and impact values of notched materials can be low (a factor to be considered before threading a product, for example).

(7) *Conductivity.* Unmodified resins are usually good insulators of both thermal and electrical conductivity. However, conductivity can be induced or enhanced by the addition of inorganic compounds, with significant variation in the final properties.

Environmental Resistance

Plastic materials do not corrode by electrochemical mechanisms, as do metals and alloys. The degradation of the material or its properties is related to the structural similarity between the material and the environmental and/or to the permeability of the material. Permeability is controlled by the solubility of the material in the environment and by the rate of diffusion of the environment into the plastic (primarily under temperature control). When diffusion into the plastic matrix occurs and the environment is chemically incompatible, all mechanical properties of the plastic are downgraded. As previously noted, a simple hardness test will often give an indication of the rate and extent of degradation. Specific comments about chemical resistance of types of plastics are given in "Specific Materials," on the following page.

Fabrication

Most of the thermoplastic materials can be hot-fusion welded, or solvent welded, at properly designed joints. Threaded couplings can be used for many materials, with careful attention to notch sensitivity. Solvent-welding is usually preferred for joining all but the polyolefins (polymers of ethylene, propylene, and butylene) for which fusion-welding is normally employed.

Specifications

There are a large number of ASTM specifications covering

materials, qualification tests, test procedures, and properties. Commodity standards and military (MIL) specifications are also relevant.

Specific Materials

(1) *Polyethylene.* The ethylene polymer (PE) is the most widely used plastic today, and is available in the form of gaskets, solid drums or containers, drum liners, piping, and small molded parts for process equipment. PE can be joined by thermal welding. PE is widely used up to approximately 60°C (140°F), especially for water service applications. It is also available as fiber-reinforced product for improved strength. *Ultrahigh density polyethylene* (UDPE) has improved mechanical properties, greater thermal stability and improved solvent resistance compared with the older type of formulations. It is attacked by strong oxidants and is susceptible to environmental stress cracking (ESC) by some inorganic (48% hydrofluoric acid) and organic chemicals (notably detergents, wetting agents, alcohols, ketones, aldehydes, and organic acids). Hydrocarbons (both aliphatic and aromatic), chlorinated solvents, and strong oxidants are to be avoided. A black-pigmented version is required to resist UV degradation.

(2) *Polypropylene.* Polypropylene (PP) is the next higher member of the polyolefin family. Stronger and somewhat more resistant chemically than PE, it is used both as a plastic liner for steel pipe and as solid pipe or sheet material. Its strength and weldability make it a prime candidate for piping for corrosive waters, dilute acids and many waste waters (provided there are no incompatible organic species contained therein). However, it is suscep-

tible to ESC in hot brine solutions and in 98% sulfuric acid, unless copolymerized with other materials.

(3) *ABS*. Acrylonitrile-butadiene-styrene (ABS) is one of the older standbys, particularly for piping systems. This material can have considerable variation in properties, depending on the ratio of acrylonitrile to the other components. Higher strength, better toughness, greater dimensional stability and other properties can be obtained, although sometimes at the expense of other characteristics. Although the ABS material has poor heat tolerance—approximately 90°C (195°F)—with relatively low strength and limited chemical resistance, its low price and ease of joining and fabrication make it attractive for distribution piping (for gas, water and waste), vent lines, automotive parts, and other consumer items. The ABS plastic will withstand attack by only a few organic compounds, is attacked by oxidizing agents and strong acids, and will stress-crack in the presence of certain organic species. Problems with ultraviolet (UV) degradation have been largely overcome through the use of protective additives.

(4) *CAB*. Cellulose acetate-butyrate (CAB) is one of a group comprising acetate, propionate, and butyrate plastics. These are used for piping and other industrial applications, although not in major quantities. They are found in a number of small items, such as name-plates, electrical component cases, high impact lenses, and other applications requiring a transparent plastic with good impact properties. Overall chemical resistance is comparable to most of the other thermoplastics.

(5) *PVC.* Polyvinyl chloride (PVC) piping is very popular for water service, particularly in the presence of chlorine or other oxidants. A variant, CPVC (chlorinated PVC), is used for *hot-water* services. PVC and CPVC have excellent resistance to both acids and alkalis, but their solvent resistance is very limited. Straight-chain hydrocarbons (i.e., aliphatics like kerosene or oil) do not affect them. Piping may be joined by hot-air welding or solvent-welding. Sheet-lining of vessels is a common application.

   PVC Type I is an unplasticized form with good chemical resistance. Type II has been modified by the addition of styrene-butadiene rubber, improving notch toughness and impact strength at some sacrifice in chemical resistance.

(6) *PVDC.* Polyvinylidene chloride. Saran[†] (PVDC) is a variant of PVC, having both chlorine atoms on the same end of the monomer (instead of opposite ends, as with vinyl chloride). PVDC has improved strength, hardness, and chemical resistance, especially to organic solvents, mineral acids, and oxidants. Formerly, a major market was in Saran-lined steel pipe. Currently, it is more often found in valves, pumps, and piping.

(7) *Fluorocarbons.* These are the most versatile and important group of plastics for the process industries. Maximum chemical resistance and heat stability are provided by this class of materials as pipe liners, tank linings, impellers, mixers, spargers, tower packing, lined valves, etc. The original polytetrafluoroethylene (PTFE) provides adequate heat stability to 260°C (500°F) for many applications, and resistance to essentially all chemicals except molten alkalis. Fabrication of PTFE is very difficult, as it

† Trade name

cannot be welded nor can it be applied adhesively except by a special pretreatment, precluding its use as a liner. Note that the term Teflon[†] no longer applies only to PTFE and should not be used without further designation.

Modifications of the original PTFE-developed rapidly. The modified variants have somewhat lower heat stability limits and slightly reduced chemical resistance, but are much more amenable to fabrication by welding, hot-forming, cold-working, and adhesion. Substantially impermeable linings or coatings, which cannot be made with PTFE, can be produced from some of these modifications. "Spaghetti-type" heat exchangers are also available. Among these variants are FEP (fluorinated ethylene-propylene), CTFE (chlorotrifluoroethylene), PVF (polyvinyl fluoride), PVDF (polyvinylidene fluoride), ECTFE (ethylene chlorotrifluoroethylene) and PFA (perfluoroalkoxy). Each has specific properties and characteristics, and special source books and manufacturers' literature should be consulted. A proper understanding of the available products is necessary to assure selection of the optimum material for a specific application.

(8) *Nylon.* Nylon is a polyamide-type plastic used primarily for mechanical parts in appliances, but nylon coatings can be applied by fluidized bed techniques. The material is heat stable to approximately 175°C (345°F), but has very limited chemical resistance.

(9) *Acetals.* The acetal resins are competitive with nylon in mechanical properties, and greatly superior in chemical resistance. Proprietary resins like Delrin[†] and Celcon[†] have three times the dimensional stability of nylon.

(10) *Polycarbonates.* The polycarbonate resins (e.g., Lexan[†]) have mechanical properties equivalent to the acetals. Their usefulness in industrial applications, however, is limited by their susceptibility to environmental cracking in even mild atmospheric exposure.

(11) *Others.* Specific useful properties can be found in such other, less frequently used plastics as polysulfones, acrylics, styrenes, and polyphenylene oxide (PPO) or sulfide. (A new thermoplastic, polybenzimidazole [PBI], has outstanding heat resistance, withstanding temperatures as high as 430°C [806°F] and in short duration, 760°C [1,400°F].) It is finding application. (In seats and seals for valves in the CPI and for high-temperature seals, packings and insulators in oil-field applications.)

18.1.2 Thermosetting Resins

This group of plastics is hard and often brittle at ambient temperatures. At higher temperatures, some loss in properties may occur, but in general these are retained until the materials degrade (usually by carbonization or oxidation). For engineering applications, all the thermosetting resins are reinforced by some means.

Reinforcement

Although silica, asbestos, paper, linen, and other plastics and materials are added as reinforcing materials, glass fiber or cloth is by far the most popular. Glass contents varying from 25 to 80% are used in the form of woven cloth and random veil, but the greatest strength is achieved by glass filaments overlaid at the proper angle. Other benefits are improved modulus and a much lower coefficient of thermal expansion than the unreinforced resin.

Fillers

Various other siliceous materials, as well as carbon and metal powders or fibers, can be added to produce even greater modulus

values, greater strength, differing thermal and/or electrical properties, and desired colors in the composite products.

Physical and Mechanical Properties.

With the almost infinite number of materials combinations
possible, it is obvious that specific properties of almost any type
can be designed into the product (even the strength of a carbon
steel). Heat-shields for re-entry from space, and graphite-fiber-
reinforced golf clubs, tennis racquets, and fishing rods are examples of such specific applications.

Fabrication

The reinforced thermosetting materials are true structural
components, which can be machined, threaded, bolted and/or glued
to fabricate equipment.

Materials

(1) *Epoxy.* Epoxy laminates, reinforced with glass
     cloth for mechanical strength, have superior
     chemical resistance. They will withstand hot
     alkalis, mineral acids up to 20% (except nitric
     acid) and most organic solvents. The tempera-
     ture limit is approximately 95°C (200°F). They
     have poor resistance to strong oxidants, amines,
     and chlorinated solvents, but epoxy-based
     paints are widely used as protective coatings
     (see Chapter 36).

(2) *Phenolics.* Phenolics are heat-curing resins
     which offer a combination of excellent proper-
     ties. Many forms of apparatus are available
     (e.g., pipe, fitting, pumps, valves, column sec-
     tions, and heat exchanger components). Phe-
     nolics are resistant to many solvents and acids,
     being adversely affected primarily by alkalis,
     oxidants and amines. Their temperature limit
     is approximately 110°C (230°F). Many types
     of process equipment use the phenolic resin for
     the inside, the exterior being "armored" with
     epoxy or polyester laminates.

They are also used in protective coatings, alone or in combination with epoxies (Chapter 36).

(3) *Polyester*. The most common designation for this type of construction is FRP (fiberglass-reinforced plastic) or GRP (glass-reinforced plastic). RTP (reinforced thermoset plastic) is now the designation used in ASTM standards and the ASME B&PV Code.

This terminology is very loose, as the polyesters comprise a family of materials which vary from ortho- and iso-phthalates to vinyl esters and bisphenol polyesters, for example. The different resins have different chemical resistance to specific environments, but in general are better than epoxies or phenolics in oxidizing situations. A large amount of tank-age, piping, duct-work, etc., is so constructed. Common applications include underground tanks for gasoline, storage tanks for mineral acids (e.g., hydrochloric, phosphoric), and piping for aggressive waters (e.g., seawater and other untreated water).

Utilization of FRP equipment in general has been limited to some extent by the lack of a proper design and construction code. In 1989, ASME issued RTP-1, "Reinforced Thermoset Plastic Corrosion Resistant Equipment," to correct this deficiency.

(4) *Furanes*. The furanes, which are derived from furfuryl alcohol, have outstanding resistance, being chemically attacked only by certain nitrogenous or chlorinated organic compounds. Their application was limited previously by problems of aging and cracking, but modern formulations and supplemental reinforcement have extended their usefulness in industry.

18.2 Rubber and Elastomers

The term *rubber* originally applied to the natural material. *Elastomer* is a term derived from the words "elastic" and "polymer," and is broadly used to embrace both natural and synthetic rubbers as well as plastics formulated to have rubber-like or elastomeric properties.

18.2.1 Specific Materials

Natural Rubber

Natural rubbers are made by processing the sap of the rubber tree, and compounding it with vulcanizing agents, antioxidants, fillers, pigments, etc. Red, black, and white rubbers are made with different pigments (e.g., iron oxide, carbon black, zinc oxide).

Natural rubbers have poor resistance to atmospheric oxygen, ozone, sunlight (ultraviolet [UV] rays), petroleum derivatives, and many organic chemicals. Resistance to non-oxidizing acids, such as hydrochloric, and alkalis is good. Rubber has been used for pumps, valves, piping, hoses, and machined products (when hardened by *vulcanization*). However, many of the historical applications have been taken over by the newer plastic materials.

Synthetic Rubbers

(1) *Buna S*. This is a copolymer of butadiene and styrene used in tires, hose, and miscellaneous goods. It offers no particular advantage over natural or other synthetic rubbers in chemical service.

(2) *Buna N*. The copolymer of butadiene and acrylonitrile, also called Hycar[†], is produced in ratios varying from 25:75 to 75:25. The manufacturer's designation should identify the percentage of acrylonitrile. Compared with natural rubber, it has good oil resistance, and is used for hose, gaskets and packing (e.g., O-rings). Chemical resistance is fair to poor generally, especially towards solvents. It has fair resistance to ozone and UV, but is severely embrittled at low temperatures.

(3) *Butyl.* Butyl rubber is a polymer of isobutylene, with small additions of isoprene or butadiene. It is remarkably impermeable to gases and is used for tire innertubes and hoses. It resists aging and ozone, and is more resistant to organic chemicals (except aromatic compounds) than are most synthetic rubbers.

(4) *Neoprene.* Neoprene is a polymer of chloroprene and was developed for resistance to oil and solvents. It is resistant to aging, ozone, and moderately elevated temperatures, and has good abrasion and wear characteristics. It is widely used for hoses, gaskets, motor belts, tires, tank linings, etc.

(5) *EPDM.* Ethylene-propylene diene methylene (EPDM or EPR) is a copolymer of ethylene and propylene. It has much of the chemical resistance of the related plastics, and has good resistance to steam and hot water, although not to oil. It has become the standard lining material for steam hoses. EPDM is widely used in chemical services as well, having a broad spectrum of resistance.

(6) *Specialty Elastomers.* There are basically three kinds of specialty synthetic rubbers:

    (a) *Silicone rubbers* are based on silicon (rather than carbon) linkages. Although chemical resistance is relatively poor, silicone rubbers withstand ageing and ozone, as well as oil, and can tolerate temperature extremes from $-75°C$ $(-103°F)$ to $200°C$ $(390°F)$.

    (b) *Chlorosulfonated polyethylene* (Hypalon[†]) has outstanding resistance to oxidizing environments. It has been used, for example, for hose to handle 96% sulfuric acid. Otherwise, its physical and chemical properties are similar to neoprene.

(c) *Fluorinated elastomers*, such as Kel-F[†] and Viton[†] will withstand oxidants, and have temperature capabilities up to 300°C (570°F) for short periods of time. However, contrary to what one might expect by analogy with the fluorinated plastics, they mostly have poor resistance in solvents or organic media. They do resist acids and alkalis. The perfluorelastomer products, such as Kalrez[†] and Chemraz[†], have chemical resistance comparable to that of the fluorinated plastics.

Super-fluorinated Viton[†] also has enhanced chemical resistance.

18.3 Other Nonmetallic Materials

Besides the plastics and elastomers, there are a number of other nonmetallic materials of interest to the process industries. These materials are used both as materials of construction in their own right and as linings, barriers, or coatings of one kind or another.

18.3.1 Carbon and Graphite

Carbon products have a homogeneous structure, which is usually produced below 1,230°C (2,250°F). Graphite is a crystalline form of carbon produced by processing at temperatures in excess of 1,980°C (3,600°F). The main use of carbon and graphite per se is as brick for lining process vessels and as ring packing for fractionation columns. In fiber form (e.g., Grafoil[†]), graphite is used to replace asbestos in gaskets, because of high heat resistance, excellent chemical resistance and outstanding sealing characteristics.

There is an appreciable difference in corrosion resistance. Within certain limits, carbon will withstand more powerful oxidizing agents, such as hot concentrated sulfuric acid or halogens (which will attack graphite rapidly). Both forms will withstand mineral and organic acids, alkalis, salts, organic solvents, etc.

Where thermal conductivity is required (e.g., for heat exchanger tubes), impervious graphite or Karbate[†] is manufactured by forming the desired shape from graphite, evacuating the pores, and impregnating with a resinous material (e.g., a phenolic resin). The impregnation seals the porosity but limits both the corrosion resistance and the maximum service temperature. New developments include impregnation with PTFE or carbon. The latter treatment permits operation to 400°C (750°F) maximum.

In addition to heat exchangers, whose thermal conductivity is close to that of copper-tubed items, impervious graphite pumps and valves are available and find broad application in process services. They are, of course, susceptible to brittle failure by mechanical shock.

18.3.2 Glass

"Glass equipment" is of two types: solid glass and glass-coated steel. In glass-lined (or externally glass-coated) equipment, one may purchase a *crystallizedglass* (e.g., Pyroceram[†]) with outstanding resistance to thermal shock, which is a potential problem with conventional glass. Crystallized glass can be heated to at least 800°C (1,470°F) and quenched in ice-water without cracking. All glasses, however, are substantially equivalent in corrosion resistance and susceptibility to *mechanical* damage.

Soft glass was formerly used in laboratory apparatus, but has been replaced with the tougher and stronger Pyrex[†] glass. Pyrex[†] glass is available also as pipe, distillation columns, and heat exchangers. Glass-lined equipment is available in the form of tanks, vessels, piping, valves, agitators, and pumps. The normal service temperature limitation is 175°C (345°F).

Glass will resist mineral and organic acids, except hydrofluoric and phosphoric acid containing fluoride contaminants, up to the atmospheric boiling point. Above those temperatures, measurable dissolution will occur. Strong alkalis (e.g., sodium or potassium hydroxides as opposed to ammonium hydroxide) will attack glass, although some glass formulations will withstand a pH of 12.

Note that some "glass-lined" vessels such as *beer tanks* are only intended to prevent iron contaminations; they are not capable of withstanding truly corrosive liquids.

18.3.3 Ceramics

Ceramics include not only the traditional silicate-based materials but also metallic oxides, borides, nitrides, carbides, etc.

These are formed into useful shapes, and given permanence and durability, by heating to high temperatures.

In the process industries, the major applications are as chemical stoneware, porcelain, and acid-brick. Stoneware and porcelain differ in appearance and porosity. They are attacked by hydrofluoric acid, fluoride-bearing phosphoric acid and alkalis, in a manner analogous to glass. However, a special proprietary grade, Body 97[†], will resist up to 20% caustic to approximately 100°C (212°F).

Ceramics are brittle materials, much stronger in compression than in tension, and must be protected against both thermal and mechanical shock. An important exception is a high-alumina (i.e., aluminum oxide) material, such as Alite[†], which can be used up to 1,000°C (1,830°F) and is more shock-resistant than other stoneware. In general, ceramics will tolerate more rapid cooling than heating, and start-up procedures can be very critical in this regard.

Acid-brick is used for lining vessels in many severely corrosive services. The brick is susceptible to penetration through its pores and at the joints. Its primary function is to reduce the temperature at which the environment encounters the substrate. With rare exceptions, it should *not* be used without a membrane (of rubber, plastic, or lead) to protect the metallic substrate, which is usually steel and low in resistance to the environment.

An important development is that of thermally conductive, highly corrosion-resistant materials in the form of heat-exchanger tubing. Hexoloy[†] (alpha-silicon carbide) heat-exchanger has a thermal conductivity close to that of copper and resists many corrosives inimical even to tantalum.

18.3.4 Refractories

Most refractories employed in process industries are castables, which comprise an aggregate (e.g., brick, shale, alumina, etc.) in a binder such as calcium aluminate, silicate, etc. The refractories are highly heat-resistant. They are classified as acidic, basic or neutral, the classification being indicative of their corrosion resistance. Acidic materials, like silica (i.e., silicon dioxide), are suitable when the environment is not *alkaline* (i.e., when it is neutral or acidic). Alkaline environments require alkaline refractories, such as alumina, zirconia, or magnesite.

One should always consider the possible presence of *sodium salts* in furnace applications because these can be oxidized to sodium oxide, with attendant attack on an acid refractory.

18.3.5 Concrete

Conventional concrete is a mixture of portland cement with an inert aggregate, such as sand or gravel. The cement itself is a mixture of calcium salts (e.g., silicates, aluminates, carbonates), which tends to resist most natural environments, such as air or water.

There are two major problems. One is the corrosion of the steel reinforcing bars, commonly known as *rebar*; the second is corrosion of the concrete itself. Internal rusting of the rebar can occur within the concrete itself, due to ingress of moisture, salts, and oxygen through the pores of the material. Since rust occupies approximately seven times the volume of the corroded steel, great internal pressures are developed, which tend to crack or spall the concrete. Rebar corrosion may be minimized by coatings (e.g., epoxy or galvanizing), sealing the concrete against ingress of corrosive species, or (with suitable design) by cathodic protection of the steel rebar within the structure.

There are three basic mechanisms for the corrosion of concrete formulated with portland cement:

(1) Corrosion I is the leaching of free lime (e.g., by soft waters or carbonic acid), or direct acid attack upon the calcium carbonate constituent.

(2) Corrosion II is a base-exchange reaction between the readily soluble components of the acid and an aggressive solution. The reaction products may either be leached out or remain in place in a *non-binding* form (e.g., magnesium ions will react to form insoluble salts, leaving a gelatinous mass in the voids).

(3) Corrosion III is the formation of new internal salts, whose crystallization and expansion destroy the concrete by internal stresses. Sulfate ions, even from neutral salts, react with tricalcium aluminate to form calcium

sulphoaluminate hydrate (the so-called sulfate bacillus) with a large crystallized water content (i.e., 31 mols of water/mol of salt).

Concrete is also attacked by certain organic solvents that react with calcium (e.g., ethyl acetoacetate and other esters).

Sulfate attack is more of a problem with flowing waters than under stagnant conditions or seeping water. Ammonium, magnesium, sodium, and calcium sulfates are particularly destructive. The first level of sulfate resistance is achieved by limiting the permissible concentration of tricalcium aluminate, but this effects only a nominal improvement. The suggested sulfate concentration limits for ASTM C 150[1] portland cements under flowing conditions are 150 ppm, 500 ppm, and 1,500 ppm for Types I, II, and V, respectively.

The adverse affect of sulfates is ameliorated by increasing chloride concentration in the water. Although quantitative parameters have not been established, it is well-known that Type I cement is entirely adequate for seawater, despite the high sulfate content (approximately 4,000 ppm) of that environment.

For more severe conditions, the cement can be pretreated with sulfite liquors or gypsum solutions to form the hydrated salts *in situ* while the cement is still in the plastic state (e.g., supersulfated cement). However, trass or pozzolanic cements (which give amorphous silicic hydrates in the structure) are more reliable (e.g., ASTM C 595,[2] Types IP and P). In extreme cases, special alumina-based cements are used.

A special warning is in order as regards use of concrete in wastewaters, where sulfates may be present, yet unanticipated. Also, hydrogen sulfide (from bacterial action or other sources) can accumulate in the vapor zone of concrete sewage pipe, oxidizing to sulfurous or sulfuric acid and causing direct acid attack.

18.3.6 Wood

Wooden tanks and piping are often an economical substitute for alloy construction, particularly in aqueous solutions containing acid salts. The cellulose component in wood may be oxidized or hydrolized by acid conditions, while the lignin is subject to attack by alkalis. Wooden construction is generally not suitable for alkaline solutions, although a pH of up to 11 may be tolerated.

Strong acids like sulfuric, hydrochloric, and nitric will oxidize, hydrolyze, or dehydrate wood. Anhydrous hydrogen fluoride will char wood instaneously. Oxidizing agents, such as nitric or chromic acid, permanganates, and hypochlorites, should not be used with wood. However, dilute acids and acid-salt solutions are handled well by wooden apparatus.

Several types of cedar, cypress, fir, redwoods, and pine have been used. Douglas fir is preferred for the more severe services, for example 95°C (205°F) in the pH range of 2 to 11, while redwood is used to 80°C (175°F) from pH 4 to 9. Heartwood is preferred for optimum resistance, while sapwood may be used where the process liquid itself actually preserves the wood (e.g., in brining or pickling foodstuffs).

Wood may be impregnated with preservatives to protect it from decay. It may also be married to plastics to make composite materials for special purposes. For example, small blocks of end-grained balsa wood can be held together with a fabric scrim or netting to serve as core for FRP to increase stiffness in large FRP tanks (e.g., horizontal tanks).

For corrosive and/or abrasive services particularly, professional assistance should be sought in selecting the particular wood and the applicable specifications.

References

1. ASTM Standard C 150, "Specification for Portland Cement" (Philadelphia, PA: ASTM).
2. ASTM Standard C 595, "Specification for Blended Hydraulic Cements" (Philadelphia, PA: ASTM).

Suggested Resource Information

ASM Engineered Materials Handbook, Vol. 4, *Ceramics and Glasses* (Materials Park, OH: ASM International, 1991).

Managing Corrosion with Plastics (Houston, TX: NACE International).

J. H. Mallinson, *Corrosion-Resistant Plastic Composites in Chemical Plant Design* (New York, NY: Marcel Dekker, 1989).

B. J. Moniz, W. I., Pollock, eds., *Process Industries Corrosion–The Theory and Practice* (Houston, TX: NACEInternational, 1986), pp. 573, 577, 589, 639, 649, 661, 681, 695, 703, and 711.

SECTION IV

Corrosive Environments

19

Corrosion by Water and Steam

In the chapters concerning corrosion by specific environments, corrosion by water (and its contaminants) will be discussed first because the presence of water is a necessary condition for atmospheric corrosion and corrosion by soils and salts. Also, water is the most commonly encountered corrodent in both municipal and industrial applications and can be a major problem in any processes especially in its role as coolant.

Water is something that everybody uses without thinking very much about it; it is something that comes out of a tap or line to be used as needed. The professional water treatment specialist understands the true complexity of the subject, but *anyone* involved in corrosion/materials work should learn at least the fundamentals, if only to know when to get professional help.

Pure water is not really corrosive, except to anodic metals above at least 200°C (392°F). The corrosive action so often encountered at lower temperatures is due to gases and minerals dissolved in the water, plus the effects of bacteria and biomasses in some cases.

19.1 Water Chemistry

In order to evaluate properly the characteristics of a water proposed for some specific use, a complete analysis must be available. Table 19.1 (on the following page) shows the basic minimum information which must be obtained and the units in which it is expressed to facilitate evaluation.

For a water sample to be truly representative for analysis, the sample bottle must be filled to the brim and tightly capped (to prevent loss of volatile species, such as carbon dioxide, or ingress of gaseous contaminants, such as air). The analytical data obtained can be used to diagnose the scaling characteristics and probable corrosivity of water.

Table 19.1

Characteristics Measured in Basic Water Analysis

Variable Measured	Units
TDS (Total Dissolved Solids)[A]	ppm (parts per million)
Ca (Calcium Hardness)[A]	ppm as $CaCO_3$
Mg (Magnesium Hardness)	ppm as $CaCO_3$
TH (Total Hardness; Ca plus Mg)	ppm as $CaCO_3$
PA (Phenolphthalein Alkalinity)	ppm as $CaCO_3$
MOA (Methyl Orange Alkalinity)[A]	ppm as $CaCO_3$
pH[A]	pH
Chlorides	ppm Cl^-
Sulfates	ppm SO_4^{2-}
Silica	ppm SiO_2
Iron	ppm Fe
DO (Dissolved Oxygen)	ppm O_2

[A] Indicates use in scaling calculations; see previous page.

19.2 Scaling Indices

Calcium carbonate has *decreasing* solubility with rising temperature and/or increasing pH. Table 19.2 provides the data from which scaling characteristics can be estimated from the analytical data for pH, TDS, Ca, and MOA. Using these data, one can calculate *either* the pH_S (pH of saturation of calcium carbonate) or the *temperature* at which scaling will occur at the *observed* pH.

The scale usually consists of calcareous deposits (mixtures of calcium carbonates and magnesium hydroxides). The calculations are based on Ryznar's modification of the *Langelier index*. In current terminology, the relationships are:

$$LSI = pH - pH_S \qquad (1)$$

where LSI stands for Langelier saturation index, and

$$RSI = 2\,pH_S - pH \qquad (2)$$

where RSI stands for Ryznar stability index.

These relationships, together with the calculation of pH_S, are given in Table 19.2. These calculations are only approximations but are often useful.

Table 19.2

Scaling by Water[1]

$pH_S = (9.3 + A + B) - (C + D)$

$LSI = pH - pH_S$

where O indicates balanced conditions;
+ indicates scale formation;
− indicates scale dissolution; and

$RSI = 2\,pH_S - pH$

where values > 7.5 indicate scale dissolution

Total Solids	A	Ca as CaCO$_3$	C	MO Alk	D
50-400	0.1	10-11	0.6	10-11	1.0
400-1000	0.2	12-13	0.7	12-13	1.1
		14-17	0.8	14-17	1.2
Temp. °F[A]	**B**	18-22	0.9	18-22	1.3
		23-27	1.0	23-27	1.4
32-34	2.6	28-34	1.1	28-35	1.5
35-42	2.5	35-43	1.2	36-44	1.6
44-48	2.4	44-55	1.3	45-55	1.7
50-56	2.3	56-69	1.4	56-69	1.8
58-62	2.2	70-87	1.5	70-88	1.9
64-70	2.1	88-110	1.6	89-110	2.0
72-80	2.0	111-138	1.7	111-139	2.1
82-88	1.9	139-174	1.8	140-176	2.2
90-98	1.8	175-220	1.9	177-220	2.3
100-110	1.7	230-270	2.0	230-270	2.4
112-122	1.6	280-340	2.1	280-340	2.5
124-132	1.5	350-430	2.2	360-440	2.6
134-146	1.4	440-550	2.3	450-550	2.7
148-160	1.3	560-690	2.4	560-690	2.8
162-178	1.2	700-870	2.5	700-880	2.9
178-194	1.1	880-1,000	2.6	890-1,000	3.0
194-210	1.0	1,000-1,200	2.7	1,000-1,200	3.1

[A] °C = 5/9 × (°F − 32).

Note: Use *actual* pH of water when calculating B, the temperature at which the water will scale.

[1] *Corrosion Basics–An Introduction* (Houston, TX: NACE, 1984).

A much more precise calculation can be made on a spreadsheet, using the formulas recently promulgated by the American Water Works Association (AWWA) in the March 1992 issue of *Opflow*.[1] The calculations take into account ionic strength, activity coefficients, etc. Once the spreadsheet entries are completed , both LSI and RSI are instantly recalculated when one or more of the parameters are changed. (These and other indices are explained in detail in AWWA publication, *Lead Control Strategies*.[2])

If the LSI is *positive*, the water is saturated with calcium carbonate at the observed temperature and pH and is therefore prone to scaling. A *negative* LSI shows a tendency to *dissolve* scale (or corrode cement or concrete or other calcareous material). It should be noted that the LSI has little to do with corrosion of metals (although calcareous deposits *may* be protective), because of occluded salts and variations in permeability to dissolved oxygen. Bacterial action under scale can also affect corrosion.

The RSI is simply a more meaningful refinement of the LSI. Values below 6.0 indicate increasing scaling tendencies. The LSI and RSI techniques are applicable to potable waters (see further below) and other fresh waters.

For recirculated cooling waters, in which a good deal of buffering occurs at pH >7.5, a more accurate calculation may be the Predictable Saturation Index (PSI). This is a variant on the RSI, the equilibrium "pH_{eq}" being substituted for actual pH in the Ryznar Index format. The pH_{eq} is read off a table which relates it to the total alkalinity (or calculated as $pH_{eq} = 1.465$ log Total Acidity + 4.54).

Two other indices of interest are the Stiff-Davis Index or SDI, which extends the usefulness of this type of calculation to oilfield brines and other high salinity waters, and the Aggressiveness Index or AI, which predicts corrosivity to cement from the pH, calcium hardness and methyl orange alkalinity (MOA), where AI = pH + Log (Ca x MOA), with values below 12 becoming increasingly aggressive to cement.

Note again that, at the measured pH, one can calculate the temperature at which the water begins to scale, by solving for the value "B" in the equation provided.

19.3 Corrosion of Steel by Water

In a general way, corrosion of iron and steel is proportional to the chloride content when dissolved oxygen (DO) is constant, and vice versa.

Corrosion of steel (cast iron is slightly more resistant in most instances) increases with chloride content of the water to reach a maximum at approximately 6,000 ppm (Figure 19.1). Above that level, the chloride effect is offset by diminishing solubility of DO, as in seawater or brines.

On the other hand, with fresh waters particularly, the corrosion of steel is governed by DO over a broad pH range (e.g., pH 4.5 to 9.5) at relatively low temperatures (Figure 19.2). Note that corrosion rate in this figure is actually expressed as mils per year *per milliliter of DO per liter*. Below pH 4.5, the corrosion is controlled by hydrogen evolution under acid conditions; above 9.5, corrosion is suppressed by an insoluble film of ferric hydroxide. In some instances, where acidity is high without a corresponding drop in pH (as with carbonic and other weakly ionized acids), corrosion with hydrogen evolution may occur in the pH 5.0 to 5.5 range.

Historically, in an *open* container, corrosion reportedly reached a maximum at approximately 80°C (175°F) because of the decreasing solubility of DO (Figure 19.3). More recent investigations, using quartz rather than glass containers, indicates that the

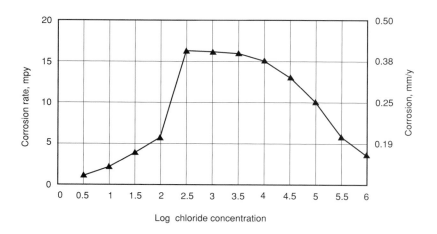

Figure 19.1 Corrosion of Steel vs Chlorides.

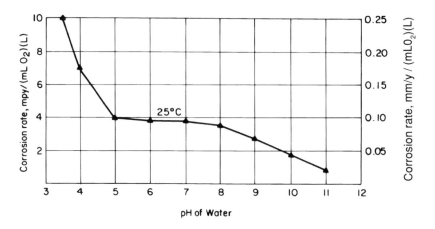

Figure 19.2 Steel: pH vs corrosion.

diminution of attack reported was enhanced by silicate inhibition resulting from some corrosion of the glass container. Note that, in a *closed* container, corrosion will increase with temperature more or less indefinitely because of retention of small amounts of DO under pressure. As we will discuss further below, this is the main basis for removal of DO from boiler feedwater.

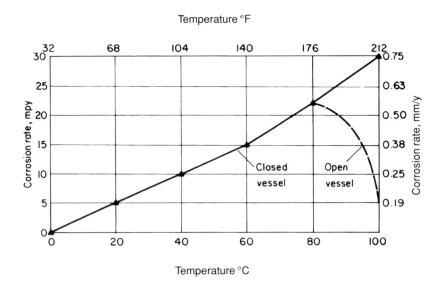

Figure 19.3 Steel: corrosion vs temperature.

19.4 Types of Waters

Waters may be classified in terms of origin, nature, and/or utilization. Many corrosion and materials problems arise from a basic misunderstanding of nomenclature and characteristics.

19.4.1 Origin

Waters may be classified as of surface or subsurface origin. *Surface waters* comprise the rivers, lakes, and seas which are the major source of water for industrial usage. *Subsurface waters* consist of waters from springs or wells and "produced water." Spring water comes to the surface from an aquifer, lifted by hydrostatic pressure. Well waters are brought forth by digging to a source. Artesian wells are those in which impermeable geological strata are perforated, allowing the water to surface by its own pressure. Other wells simply pump the water to the surface, usually using submerged pumps. Produced water is peculiar to oil- and gas-well drilling operations, being coproduced with the hydro-carbons. It may be fossil water, entrapped for eons in the geological strata before being released by the drilling operation (as distinct from an aquifer which is replenished by drainage from surface sources or rainfall).

19.4.2 Natural Waters

Natural waters particularly are classified as *fresh, brackish,* or *salt* (e.g., seawater). Fresh water is also classified as hard or soft, although these are relative terms on a sliding scale and should be quantified according to the LSI or RSI. Roughly, water considered soft if the hardness is in the 0 to 75 ppm $CaCO_3$ range. Moderately hard water contains 75 to 150 ppm and hard water >150 ppm calcium hardness.

(1) Fresh water typically contains less than 1,000 ppm of chloride ion.

    (a) Soft water is low in calcium and magnesium salts, lathers easily with soap and is difficult to rinse.

    (b) Hard water is relatively high in calcium and magnesium and tends to form insoluble curds when used with conventional sodium-based soaps. Originally, water hardness was actually determined by titration with standardized soap solutions, but these have been replaced with more modern analytical techniques (e.g., titration with ethylene diamine tetracetate; EDTA).

(2) Brackish waters typically contain more chlorides than fresh water, but less than seawater (e.g., 1,000 to 10,000 ppm). Such waters are found as surface waters due to salt deposits in the ground, as tidal waters (mixed from river and seawater), and as subsurface geological brines or saline wells.

(3) Seawater typically contains from 2.5 to 3.5% sodium chloride, depending on its geographical location. The salt content may increase due to surface evaporation, as in the Persian Gulf or the Dead Sea.

(4) Brines are still more concentrated salt solutions, either natural (as in oilfield or geothermal brines) or artificial (i.e., formulated for refrigeration purposes to take advantage of their low freezing point).

19.4.3 Use

From an industrial standpoint, waters are more often classified by their particular use. There are really three categories. The first (category I) involves nomenclature which clearly tells one something about the chemistry or nature of the water. In category II, there is an *implicit* (but not wholly reliable) statement about the nature of the water. In category III, the common name says really nothing about the nature or properties of the water despite the popular jargon.

Category I

Distilled water has been boiled and the vapors condensed specifically to separate it from its mineral salts. It is supposed to be pure and its major use is in the laboratory. Even so, the actual purity depends on the efficiency of the distillation system. Triple-distilled water has the highest purity and is used in scientific work, where required, as *reagent water*.

Demineralized water (DM) has had both cations and anions removed by the action of specific ion exchange resins, substituting hydrogen ions for the cations and hydroxyl ions for the anions. The resulting water is free of almost all its mineral content except perhaps a little silica. It will contain a trace of organic species from the synthetic resin contact. In industrial use, demineralized water is considered to be equivalent to distilled water in quality (e.g., in nuclear reactor applications; as boiler feedwater make-up for high pressure boilers). In fact, it can be contaminated with traces of material from the various operations involved, such as resin regeneration. The specific conductance of both distilled and demineralized water is a measure of purity and can be correlated with total dissolved solids (TDS). Mixed bed demineralizers can produce water of 0.055 μS). Super high-purity water is required in modern semi-conductor and computer manufacture. *Ultra high-purity* water requires continuous circulation, very careful selection of materials to avoid contamination, continuous ultrafiltration, and continuous treatment (e.g., ozonation or ultraviolet light) against biological activity.

Potable water is fresh water of restricted mineral content (e.g., less than 250 ppm chlorides by U.S. standards; less than 350 ppm by World Health Organization [WHO] standards), which has been specifically sanitized for drinking purposes by treatment with chlorine, chloramines, chlorine dioxide, ozone, or other oxidizing biocides. It is therefore free of coliform and other potentially harmful bacteria, within the control limits of the treating system.

Raw water is a term used in some parts of the world to indicate an otherwise untreated water which may have been disinfected by chlorination but exceeds the mineral content stipulated for truly potable water.

Seawater is water from the oceans and seas. Typically, it is encountered as the medium for marine transportation in various hulls and as once-through cooling water. Although its composition varies with geographical location and temperatures, it is roughly 3.5% salt, of which approximately 85% is sodium chloride and the remainder sulfates (calcium and magnesium). It also contains marine organisms and bacteria which can pose specific problems of scaling, fouling, and microbiologically influenced corrosion (MIC). There is a large volume of published literature on corrosion by seawater.

Category II

(1) *Softened water* is water from which all or most of the total hardness has been deliberately removed to prevent scaling in service. There are several processes which involve either pre-cipitation of calcium and magnesium salts or replacing those cations with sodium ions. Hot or cold lime softening diminishes the calcium and magnesium concentrations while Zeolite[†] softening replaces all cations with sodium ions (by ion exchange), raising the pH simulta-neously. Softened (and naturally soft) waters are *extremely* corrosive unless completely de-aerated, attacking even copper-based alloys.

(2) *Boiler feedwater make-up* (BFWMU) is the *new* water added to a boiler. It has been softened to some desired quality, but is not yet deaerated. It may range from partially soft-ened to fully demineralized water, depending on the pressure rating of the boiler.

(3) *Boiler feedwater* (BFW) may be the same as the make-up water or it may be largely re-turned steam condensate (or some mixture of make-up and condensate), depending on how a steam plant is operated and how much of the steam is condensed and recovered for recircu-lation. In many plants, a good figure might be 95% returned condensate and 5% make-up comprising the BFW.

(4) *Condensate*, in steam plant terminology, is condensed steam. It differs from distilled water in degree of contamination, as by dis-solved iron, oxygen, or carbon dioxide and/or

[†] Trade name

entrained caustic, sodium carbonates, chlorides and silica. Good quality condensate should have a conductance of not more than 50 micromhos.

(5) *Desalinated water* is water produced from a seawater or other saline source by certain regimens of treatment. It is not necessarily a high-quality water.

(6) *Utility water* is a term sometimes used to describe a water of too high a salinity to be potable but otherwise sanitized to a degree which permits its use in eyebaths, safety showers, etc. However, in some instances, it may even be seawater.

Category III

(1) *Process water* is any kind of water applied to or recovered from a chemical or other process. It may thereby contain strange or unusual contaminants.

(2) *Chilled water* is a water specifically subcooled for temperature control of a reaction.

(3) *Tempered water* is held at some specific elevated temperature for the same purpose.

(4) *Produced water* (or *formation water*) is the water phase from an oil or gas well, usually containing variable amounts of chloride ions, carbon dioxide, hydrogen sulfide, etc.

(5) *Fire water* (or *fire-control water*) is any kind of water used for that purpose. It varies from potable water to seawater.

(6) *Cooling water* is used to remove sensible and latent heat from plant process streams. It may be from any source and of any quality.

(7) *Waste water* may be either municipal (which may combine sanitary waste with some industrial waste) or purely industrial. Industrial

waste water may contain any and all types of organic and/or inorganic species. However, antipollution requirements are encouraging increasingly specific waste treatment facilities to protect the environment and encourage re-use of waste water for cooling or for BFWMU purposes. Sanitary waste water usually, but not always, originates as potable water, whereas municipal and industrial wastes are of indeterminate inorganic chemical composition.

19.5 Boilers, Steam and Condensate

A boiler is a device for generating steam by heating water to the boiling point at a particular pressure. The steam generated may be used for heating (e.g., as in reboilers or calandrias) or to run engines or turbines (e.g., for generation of electricity). Fossil fuel boilers use coal or oil as fuel for combustion, while nuclear boilers utilize the heat from atomic reactors to boil the water.

19.5.1 Boiler Water

In order to be nonscaling, the BFWMU must be softened or demineralized. In order to be noncorrosive to the steel and low alloy steel components from which boilers are usually constructed (nuclear plants also use steel in the secondary loops for steam), the water must be thoroughly deaerated.

The make-up water is rendered non-scaling by a means appropriate to the design and operating pressure of the boiler (e.g., Zeolite[†] softening up to perhaps 600 psi; demineralizing at 900 psi and above). It is then partially deaerated by thermomechanical means and the last vestiges of DO removed by chemical agents (e.g., sodium sulfite or hydrazine). The chemical scavengers may be catalyzed to increase the speed of reaction with DO.

Returned condensate, which is combined with BFWMU to constitute the BFW, may be treated (i.e., polished by filtration and ion exchange) to prevent accumulation of iron or other salts in the system.

In addition to these *external* treatments, the boiler water chemistry is further controlled by internal treatments to adjust the pH and control solids precipitated in the "mud drum." An alkaline pH of the order of 9 to 10 is usually desired.

Coagulation

In boilers operating below 250 psi (1.7 MPa) with an unsoftened high-hardness feedwater, sodium carbonate and/or sodium hydroxide supplement the natural alkalinity. The carbonate causes deliberate precipitation of calcium carbonate, magnesium hydroxide, and magnesium silicate under controlled conditions to prevent scale.

Phosphate

This approach is used in boilers below 250 psi (1.7 MPa) with a soft make-up and above 250 psi where high sludge concentrations are undesirable. The feedwater hardness has to be limited to 60 mg/L maximum, and the natural alkalinity may be supplemented by caustic additions. A sodium phosphate compound addition effects precipitation of calcium, magnesium, and silicon compounds.

Coordinated Phosphate

Specific ratios of sodium diphosphate to trisodium phosphate can effect pH control within the boiler drum without the presence of the potentially harmful hydroxyl ion (OH^-) as well as reacting with any calcium incursions. A reliably pure and consistent BFW must be available. Sludge-conditioning agents (e.g., tannins, lignins, anionic carboxylates) are also added, depending upon specific boiler operating conditions, which prevent the formation of large crystalline deposits.

Chelant

The sodium salts of ethylene diamine tetracetate (EDTA) and nitrilotriacetic acid (NTA) form *soluble* complex ions with calcium and magnesium. This helps to minimize blowdown, as described further below. Such treatment is more expensive than phosphate-type additions and is limited to boilers operating below approximately 1,500 psi (10 MPa). Polymer additions for scale control also increase the effectiveness of chelation programs.

Regardless of the degree of external softening or demineralization, enough solids will eventually accumulate in the mud drum to pose operating problems unless it is subjected to *blowdown*(BD). This continuous or intermittent blowdown controls the residual solids. (*Note:* The terms make-up and blowdown are also be encountered in the discussion of recirculated water systems of other kinds.)

A simplified diagram of a two-drum boiler is shown in Figure 19.4.

19.5.2 Steam

Dry steam is noncorrosive to steel below approximately 370°C (700°F). Wet steam, or mixtures of steam and water as with the watertube of a boiler, is likewise noncorrosive in the absence of contaminants. However, it is not uncommon to experience entrainment of water salts or boiler-treatment chemicals. Sodium salts may react with oxide films or with the metal itself to liberate atomic hydrogen under certain conditions. More often, they simply cause *caustic gouging* or *caustic embrittlement* (see Chapters 8 and 27).

Silica (i.e., silicon dioxide) is quite volatile and may be carried over in high pressure steam (unless properly controlled) to form deposits on steam-driven turbines, for example, adversely affecting their balance.

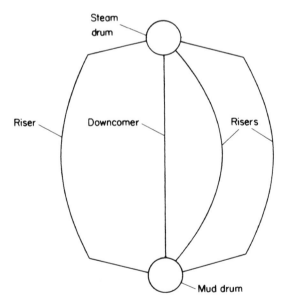

Figure 19.4 Diagram of two-drum boiler.

19.5.3 Condensate

The steam condensate from a properly controlled boiler should likewise be noncorrosive. Unfortunately, it may be contaminated by carbon dioxide (e.g., from a high bicarbonate make-up water) or DO (e.g., by ingress to the condensate return; by improper deaeration of BFW). Corrosion in steam condensate systems is usually controlled, if required, through additions of either neutralizing or filming amines (e.g., morpholine, octadecylamine, etc.).

19.6 Cooling-Water Systems

There are basically two kinds of cooling water systems, once-through and recirculated. The former is employed where there is an abundant source of surface water, but is falling into disfavor as the effects of *thermal pollution* are recognized. (Thermal pollution is the potentially adverse effect of relatively high-temperature discharged cooling water on aquatic life.)

19.6.1 Once-Through Systems

Where there is an adequate supply of inexpensive raw water, industries may elect simply to pump water through the plant heat exchangers and return it to the source. This is permissible if thermal pollution is not a problem and if there is adequate pollution control from the chemical standpoint.

Usually, the water source must be presumed to be corrosive since the surface waters are open to the atmosphere. In the past, some rivers had no DO (because of municipal or industrial pollution), but this condition does not now apply. It must be remembered that both fresh and saline waters are sufficiently corrosive to steel that more corrosion-resistant materials must be employed. The volume of water circulated through such a system effectively precludes chemical inhibition from a cost-effective standpoint.

In fresh water, galvanized steel is often adequate but brackish or salt water requires more resistant materials (e.g., copper alloys; superstainless steels; plastic, FRP, or plastic-lined steel; concrete, etc.). Table 19.3 (on following page) suggests recommended materials of construction for a variety of equipment in both fresh and seawater once-through systems.

Table 19.3

Materials of Construction for Once-Through Systems

Equipment	Fresh Water	Seawater
Screens	Coated steel Alloy 400	FRP or coated cast iron body, alloy 400 screens
Filters	FRP or coated steel; Cu or alloy 400 elements	Same
Pumps	Stainless (SS)[A], Bronze, Ni- Resist† with SS impeller duplex or austenitic alloys	Ni-Resist† with alloy 400 impeller; high-performance
Chlorinators	PVC- or Saran†-lined steel	Same
Piping	Cement-lined cast iron or concrete underground; FRP, Cu or galvanized steel aboveground	Concrete underground; FRP, 90-10 Cu-Ni, duplex or 6 Mo alloys
Valves	Bronze or SS	Alloy 400, bronze or high- performance alloys

Exchangers	Seawater in Tubes Only	
Tubes	Cu, admiralty, SS alloys 625 and C-276	Ti, Cu-Ni, 6 Mo alloys,
Tubesheets	Bronze, SS	Ti, Ti-clad, bronze, 6 Mo
Baffles	Steel,[B] bronze, SS	Same metal as shell
Tie-rods	Steel,[B] bronze, SS	Same metal as shell
Shell	Steel,[B] Cu, SS, FRP	Metal[C] or FRP
Heads	FRP, bronze, lined steel	FRP or 90-10 CuNi

[A] Stainless Steels.
[B] Only with water in tubes and steel compatible with process side.
[C] Any metal/alloy compatible with process.

In once-through systems, chlorination is usually necessary to control biological growth (e.g., bacteria, slime, marine organisms).

19.6.2 Recirculated Systems

When water is in short supply or when its chemistry must be rigorously controlled, recirculated systems are employed. These may be considered in two categories; closed and open.

Closed Recirculated Systems

In a closed recirculated system, there is no make-up after the initial charge (except to replace accidental leakage) and no blow-down. Since there is no opportunity for evaporation either, the water chemistry can be established and easily and inexpensively maintained at the initial charge. These systems are exemplified by the automobile radiator and by engine jacket cooling systems.

These systems may be treated either by rendering them sterile and anaerobic or by use of either oxidizing or non-oxidizing inhibitors. The cost of treatment, including softening and pH control, if required or desirable, is minimal.

These systems may be treated either by rendering them sterile and anaerobic or by use of either oxidizing or non-oxidizing inhibitors. The cost of treatment, including softening and pH control, if required or desirable, is minimal.

The heat which would otherwise accumulate in the closed loop is removed either in air-cooled heat exchangers or in water-to-water exchangers cooled by an external cooling water system.

Bactericidal treatment may be required, using a *nonoxidizing* biocide, such as hexamethylene biguanide, if sulfate-reducing bacteria would otherwise be a potential problem.

Open Recirculated Systems

Open recirculated systems use ponds, fountains, or cooling towers to dissipate heat by evaporative cooling. All of these types involve constant air saturation as well as some concentration of water-borne solids in the circulating water. Such systems are inevitably corrosive to steel (unless suitably inhibited) and potentially scaling unless the hardness, pH, and alkalinity are also controlled. Fortunately, as discussed below, they are usually amenable to economical treatment, unlike the once-through system with its higher water usage, especially if cooling towers (rather than fountains or ponds) are employed.

In a cooling tower system, the process equipment is cooled by the recirculated water, and the warmed water is then cooled by being sprayed over the tower packing while air is blown or drawn through the cycle water (Figure 19.5). The total amount of water actually used is limited to that lost by evaporation plus the blowdown (BD) established to limit the buildup of salts and solids

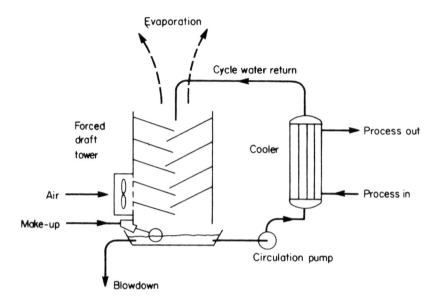

Figure 19.5 Diagram of cooling tower.

in the system. The extent of the soluble salt concentration is expressed as *cycles of concentration*, which is the ratio of chlorides in the BD to those in the make-up (MU), or the ratio of MU/BD in volume per unit time (e.g., L/s or gpm). For example, if three volumes of water were to be boiled away in a pan until only one volume of water remained, the residual water would contain three times the soluble salts of the original charge, analogous to three cycles of concentration.

The amount of water evaporated in a cycle water system is determined by the circulation rate (CR) and the heat load on the system (which is indicated by the temperature difference (ΔT) in°F[1] between the cool water from the basin or sump and the warm water returning to the tower from the plant). The evaporation rate (ER) is then:

$$ER = CR \times \Delta T/1000 \qquad (3)$$

while the cycles of concentration (C), makeup (MU) and blowdown (BD) are related as follows:

$$MU = ER + BD \qquad (4)$$

[1] To obtain SI equivalents, measure CR in liters per second, ΔT in °C, and divide by 33,330 (1 gpm = 0.06333 L/s).

$$C = \frac{MU}{BD} = \frac{ER + BD}{BD} \qquad (5)$$

$$BD = \frac{ER}{C - 1} \qquad (6)$$

Thus, for a cycle water system requiring 5,000 gallons per minute (19,000 L/min) circulation and a ΔT of 15°F (–3°C), the ER would be 75 gal/min (284 L/min). To maintain four cycles of concentration (which is a good working value), the BD = 75/3 = 25 gal/min (94 L/min) and the MU would be 100 gal/min (379 L/min). Note that this system only *uses* 100 gal/min while a once-through system would use the entire circulation rate of 5,000 gal/min (19,000 L/min). Actually, water-treatment chemicals need only be replaced in accordance with the *blowdown* rate. This feature is what makes open recirculated systems preferable when chemical treatment is required and closed systems are impractical.

In most such systems (water chemistry permitting), the optimum savings are effected at four to six cycles (Figure 19.6). The additional savings from a higher number of cycles is usually offset by the increasing difficulty of coping with higher dissolved salt and hardness concentrations.

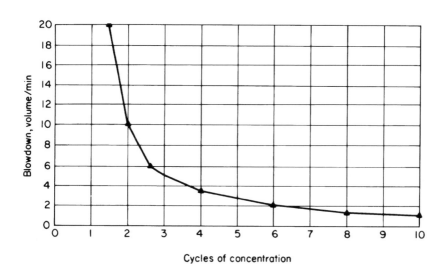

Figure 19.6 Water savings vs cycles of concentration.

Note that because of the warm temperatures and constant air scrubbing in the tower, the water is not only corrosive but a breeding ground for slime and algae introduced from air-borne spores. There is also a tendency to pick up particulate matter from the air.

Open recirculated systems must usually be corrosion-inhibited (if steel-tubed condensers are to be successfully employed; see Chapter 35), treated with biocides to control biological growths, and chemically treated to control scale and deposits. The cost of such treatment must be balanced against the obvious savings in water consumption. Such water treatment is an area in which even the competent corrosion engineer may need professional help from a specialized consultant or from a professional water-treating company.

Modern pollution control requirements are limiting some previously useful chemicals, such as chromate-based inhibitors, or requiring that they be removed from the blowdown. Fortunately, new treatments are being constantly studied and commercialized, although they may be of limited applicability in more aggressive high-chloride waters. The choice between corrosion inhibition vs resistant materials of construction is simply one of economics and pollution abatement considerations.

Table 19.4 suggests suitable materials of construction for fresh-water type open recirculated systems.

19.7 Microbiologically Influenced Corrosion (MIC)

Some allusions have already been made to microbiologically influenced corrosion, commonly referred to as MIC. Strictly speaking, MIC is not a specific form of corrosion. Rather, it consists of both direct and indirect effects on specific materials by bacteria, fungi, algae, and yeasts. Microbes can selectively consume molecular hydrogen (promoting cathodic polarization), oxidize metal ions, or produce ammonia, methane, sulfur, or hydrogen sulfide. Biomasses of one kind or another can induce localized corrosion, such as underdeposit corrosion (UDC) by concentration cells, when adhering to a metal surface. Some forms of localized corrosion, such as pitting, dealloying, IGA, and envir-

Table 19.4

Materials of Construction for Cooling Tower Systems

Equipment	Inhibited Water[A]
Filters	FRP or cast iron
Pumps	Steel or cast iron volutes, Bronze or SS[B] impellers
Chlorinators	PVC-lined or Saran†-lined steel
Piping	Carbon steel
Valves	Cast iron or steel, SS trim

Exchangers	Water on shell	Water in Tubes
Tubes	Steel, Cu, admiralty, SS, or nickel alloys	Same
Tubesheets	Steel (if process compatible); Cu, SS or Ni (solid or clad)	Same
Baffles	Steel; other	Metal compatible with process
Tie-rods	Steel; other	Metal compatible with process
Shell	Carbon steel	Metal compatible with process
Heads	Metal compatible with process	FRP, lined or coated steel

[A] Uninhibited aerated water is handled the same as once-through water (see Table 19.3).
[B] Stainless Steel.

-nmental cracking (SCC or SSC), can be caused by specific chemical agents (e.g., ammonia, sulfides, ferric, or manganic chlorides) produced by metabolic processes in, or decay of, biomasses. Any films or deposits can occlude and concentrate chloride ions on the metal surface, in larger concentrations than exist in the bulk solution.

Aerobic bacteria, such as the "iron-eating" *Gallionella* and *Crenothrix* which metabolize iron and manganese salts and excrete metallic hydrated oxides, are often associated with unexpected chloride effects. Both type 300 series stainless steel and some high-performance alloys have suffered pitting and SCC due to this effect.

More commonly, problems arise from *anaerobic bacteria* such as the sulfate-reducing bacteria (SRB) strains found in many waters and water-bearing soils. Typically, these induce hydrogen sulfide-related problems or chloride-sulfate synergistic effects. SRBs are not entirely killed by ordinary chlorination and related bactericidal treatments. Small numbers survive in a dormant condition and can establish new colonies under locally anaerobic conditions.

It is important to note that SRBs can grow under films or deposits in otherwise aerated water.

19.8 Behavior of Materials

Some generalizations are given below which characterize the behavior of engineering materials in water service.

19.8.1 Carbon and Low Alloy Steels

With the exception of steam condensate (in which an ASTM A 242[3] weathering steel may show some advantage), carbon and low alloy steels show approximately equivalent corrosion in water. Corrosion is primarily under the control of DO (or other oxidizing agents, such as chlorine or dissolved sulfur) over a broad pH range, as shown in Figure 19.2. Corrosion reaches a maximum at approximately 6,000 ppm chloride ion. Generally speaking, steel should not be used in uninhibited water systems. However, a useful life is obtained if the water is naturally anaerobic (with no bacterial action) as from artesian wells, or in essentially closed systems (e.g., fire control water storage, hot water heating). In closed systems, the water becomes self-deaerated by superficial corrosion and noncorrosive until or unless fresh DO effects entry to the system, or unless SRBs provide dissolved sulfur as an alternative corrosive agent. A Corrosion Index for steel in aerated, uninhibited water is:

$$CI = \frac{(me\ Cl^- + me\ SO_4^{2-})}{me\ M\ Alk^{(A)}} \qquad (7)$$

[A] As $CaCO_3$.

Higher CI numbers indicates higher corrosivity, other factors being equal.

19.8.2 Cast Iron

Cast iron has an inherently better resistance than steel in most natural waters because the graphite flakes tend to aid in the adherence of a more protective rust film. The resistance is still frequently inadequate for long-term service, and modern usage usually calls for internal cementitious coatings or organic films. Soft aggressive waters cause graphitic corrosion of cast iron, as do acidic waters (e.g., coal-mine runoff).

19.8.3 Zinc

Zinc has good resistance in its own right to many natural waters. However, because of its anodic nature, it corrodes preferentially in contact with steel. Galvanized steel resists fresh water attack *until* the substrate is exposed, after which the zinc coating corrodes preferentially to protect the steel. The corrosion rate of zinc is linear with time, so protection is a function of the thickness of the zinc, whether galvanized or electroplated.

Galvanized steel lacks adequate resistance in high chloride waters or seawater, and in generally unsuitable in waters of <140 ppm hardness. Further, the potential of zinc becomes cathodic to steel in some fresh waters of specific chemistry (e.g., a high bicarbonate-to-chloride ratio; Cl^- 10 ppm or less), and accelerates corrosion of the substrate. This condition is aggravated by nitrates. Accelerated pitting of some hot water systems has been ascribed to this reversal of potential and an unproven water should be tested at approximately 80°C (175°F) to determine the possibility of this phenomenon.

19.8.4 Aluminum

Aluminum and many of its alloys have good resistance in many natural waters. However, resistance is affected by water chemistry and flow conditions. High chloride waters will pit aluminum under stagnant conditions. Aluminum is also subject to attack under deposits (UDC) by oxygen cell effects and under organic matter (*poultice corrosion*). It is corroded by heavy metal salts from other equipment upstream (cementation). It has, however, good resistance to steam condensate, in the absence of appreciable iron or caustic contamination.

19.8.5 Copper Alloys

Except for dezincification of brasses containing more than 15% zinc (unless specifically inhibited), copper and its alloys are the most reliable and cost-effective alloys for many water services. Copper, red brass, admiralty brasses, aluminum brass, aluminum bronze, and cupronickels, in that order, are used for waters of increasing salinity and/or velocity. In seawater, 90-10 Cu-Ni with iron should be specified for condenser tubing in most cases where a copper alloy is desired.

Soft waters can be *highly* corrosive to copper alloys in the presence of DO, and the "super" grades of stainless steels, high-performance alloys, or plastic construction is recommended. Also, copper can suffer pitting under some conditions. Sulfide films can cause pitting in seawater, while three kinds of pitting have been described in fresh waters. Type 1 pitting is apparently caused by residual carbonaceous films from the tube manufacturing process. Type 2 is associated with hot soft waters (> 60°C [140°F]), while Type 3 pitting may occur in cold water of high pH and low salt concentrations, for unknown reasons.

19.8.6 Stainless Steels

Stainless steels will resist pure water. It is the dissolved salt that causes problems. Any 12 chromium stainless steel will resist condensate or even BFWMU, if the water is totally deaerated. The type 300 series stainless steels are prone to pitting (and SCC at moderately elevated temperatures) even in potable water under stagnant conditions, usually due to UDC or MIC, although aerated *high-velocity* seawater can be handled in type 316 (if it is kept flowing). Nominal chloride limitations are approximately <500 ppm Cl^- for type 304 and <3000 ppm for type 316L. Type 300 series stainless steel grades can be used for water-cooled condensers if the tube-wall temperature does not exceed 50°C (125°F).

Optimum reliability in water service is provided by the "superferritic" grades and by the high-performance stainless steels or the 6% molybdenum alloys described in Chapter 13.

19.8.7 Nickel Alloys

Alloy 400 (UNS N04400) can be attacked in soft waters of critical CO_2-to-DO ratios, and as heat-exchanger tubing can pit in fresh waters. However, it is widely used for long-term fresh water and many seawater applications. Alloy 625 (N06625) or C-276 (N10276) is used for seawater-cooled condensers where the process-side conditions are incompatible with copper or titanium alloys.

19.8.8 Titanium and Zirconium

Titanium is one of the most cost-effective materials for seawater applications, especially as condenser tubing. However, it has been known to pit in very hot seawater under severe conditions. Zirconium is equally good, although more expensive and used only where process-side conditions require. Zircalloys[†] (e.g., R60802) find application in water-cooled nuclear reactors.

19.8.9 Plastics and Elastomers

Within their inherent temperature and pressure limitations, the plastic materials in general are excellent for water service. PVC, CPVC, PE, PP, and others are useful in all kinds of natural waters. FRP systems based on many resin systems (e.g., epoxy, phenolic, polyester, vinyl ester) are used where still greater strength is required. Plastic-lined steel is useful where still higher strength and temperature limits are required. Elastomers are useful to approximately 80°C (175°F), with EPDM withstanding even low pressure steam.

These comments refer to natural waters, inhibited or not. In waste waters, both plastics and elastomers can suffer degradation by continually absorbing trace amounts of incompatible organic species (e.g., sorbic acid vs polypropylene; aromatic hydrocarbons or chlorinated solvents vs rubber).

19.8.10 Concrete

Portland cement-type concrete structures are suitable for water immersion if a type appropriate to the sulfate concentration, chlorides and other chemical variants is selected. Even Type I cement is suitable in seawater, as previously described.

19.9 Special Information Sources

In addition to the scientific and engineering publications which deal specifically with water, several water-treating companies (e.g., Betz, Nalco, and other companies) publish handbooks which discuss water treatment for corrosion and scale control in great detail. The technology of water treatment for industrial usage is a complex and on-going subject in which professional help is extremely useful.

References

1. *Anon., Opflow* 18, 3 (1992).
2. *Lead Control Strategies* (Denver, CO: AWWA, 1989).
3. ASTM Standard A 242, Specification for High-Strength Low-Alloy Structural Steel (Philadelphia, PA: ASTM).

Suggested Resource Information

Betz Handbook of Industrial Water Conditioning, 9th Ed. (Trevose, PA: Betz Laboratories, 1991).

C. P. Dillon, *Materials Selection for Once-Through Water Systems,* MTI Publication No. 43 (Houston, TX: NACE International, 1994).

C. P. Dillon, *Performance of Tubular Alloy Heat Exchangers in Seawater Service in the Chemical Process Industries*, MTI Publication No. 26 (Houston, TX: NACE International, 1987).

F.N. Kemmer, ed., *The Nalco Water Handbook,* 2nd Ed. (New York, NY: McGraw-Hill, 1988).

G. Kobrin, ed., *A Practical Manual on Microbiologically Influenced Corrosion* (Houston, TX: NACE International, 1993).

B. J. Moniz, W. I. Pollock, eds., *Process Industries Corrosion—The Theory and Practice* (Houston, TX: NACE International, 1986), pp. 205 and 215.

B. Todd, P. A. Lovett, *Marine Engineering Practice,* Vol. 1, Part 10, *Selecting Materials for Seawater Systems* (London, UK: Institute of Marine Engineers).

A.H. Tuthill, "Experience with Stainless Steel in Low-Chloride Waters," *Proceedings of Intl. Symposium-Materials Performance and Maintenance,* Vol. 25, Metallurgical Society of Canadian Institute of Mining, Metallurgy, and Petroleum, Symposium held August, 1991 (Elmsford, NY: Pergamon Press, 1991), p. 135.

20

Corrosion by Soil

Dry soil is not corrosive. It becomes so by virtue of the water content and related water-soluble salts which allow it to function as an electrolyte. In that event, we are dealing with an immersion condition similar to that encountered in corrosion by water. Just how corrosive a soil may be to a specific material (and we are primarily concerned with steel and cast iron) depends upon the specific constituents in the soil, its degree of aeration, and its bacterial content.

20.1 Types of Corrosion

Moisture transferred to the soil from the atmosphere contains specific contaminants (as discussed further in Chapter 21) and picks up specific water-soluble materials from the soil (e.g., salts of aluminum, calcium, magnesium, and sodium; sulfates, chlorides, carbonates, phosphates, silicates, etc.). For this reason, all of the types of corrosion to which construction materials are susceptible can be encountered.

Besides conventional phenomena and localized cells, however, buried structures of significant length (e.g., pipelines) are subject to *macrocell* action. Whole sections of line may become anodic to other long sections because of differences in soil chemistry, soil compaction, bacterial action, and thermal effects, as by gas compression heating effects or down-hole temperature gradients. As previously mentioned, stray DC currents will cause electrolysis.

In addition, fluctuations in the earth's magnetic field can induce *telluric currents*, and AC power lines can induce stray currents, both of which can cause corrosion of buried structures.

20.2 Factors in Soil Corrosion

There are three factors which influence the corrosivity of soil; resistivity, chemistry, and physical characteristics.

20.2.1 Resistivity

Resistivity is the property of a material, as opposed to *resistance* which is the resultant property of a physical entity. For example, copper has a certain resistivity but a piece of copper wire has resistance.

Soil resistivity is the net result of many chemical and physical aspects of the soil. The resistivity is determined by passing a known current, I, through a known volume of soil (volume = depth × width × length) and measuring the difference in voltage E due to current flow. By Ohms' Law,

$$R = \frac{E}{I} \tag{1}$$

and resistivity (ρ) is

$$\rho = R \times \frac{w \times d}{l} \tag{2}$$

where w is width, d is diameter and l is length. If R is expressed in ohms (Ω), area (w × d) in square centimeters, and length (l) in centimeters, resistivity must be expressed in ohm-centimeters (Ω-cm). For example, seawater has a resistivity of 25 to 75Ω-cm; only a specific *volume* of seawater has so many ohms resistance.

This concept is useful in categorizing soils because of the interaction of chemical and physical effects cited. Generally, soil is considered highly corrosive if its resistivity is 2,000 Ω-cm or less. Between 2,000 and 10,000Ω-cm, it is moderately corrosive, while at 10 to 20,000Ω-cm it is normally considered only slightly corrosive. In everyday life, CP is recommended for buried steel pipelines if the soil is below 25,000 Ω-cm. Rare instances have been cited of corrosion in 100,000 Ω-cm soil.

20.2.2 Soil Chemistry

As mentioned above, different types of soil can contain a wide variety of chemical species. Some of these will not only influence the electrolytic nature of the soil but will also have specific ion effects (e.g., chlorides for pitting or SCC; sulfides for SSC).

Varying amounts of water may be available from water-table or atmospheric ingress. DO will vary with physical condition and with effects of decaying organic material and bacterial action, as by SRBs.

Some investigators indicate that the overall redox potential (i.e., the potential which is the net result of all the possible oxidation and reduction reactions in the soil) is very important. So, of course, is the pH of the soil.

20.2.3 Composition and Condition

Soil may consist of different components, perhaps even in different strata. Sand, clay, loam, and rock are four major categories.

Any of these, with the exception of nonporous rock, can have varying degrees of moisture and chemical contaminants. Wet, salty sand is a totally different environment than dry desert sand.

Ordinary earth will also have different degrees of compaction. Freshly excavated and filled trenches will have far more accessibility to moisture and oxygen than undisturbed soil.

20.2.4 Overall Corrosivity

Rather than judging a soil solely by its resistivity, several authorities have devised tables for overall rating of soil according to the several parameters described in Table 20.1 (on the following page). This table gives different arithmetical point figures for the several parameters, which can be totaled for an overall appraisal of the soil corrosivity. The parameters usually considered include pH, chloride content, redox potential, type of soil, and soil resistivity. Point allocations are higher with the more corrosive conditions, with high cumulative scores indicating greater corrosivity (e.g., 15+ indicating severe conditions, 0 to 5 mild corrosion only).

20.3 Behavior of Specific Materials

The behavior of different materials of construction in soil will vary with the basic chemistry of the soil. We can, however, make certain generalizations and point out specific problems.

Table 20.1

Assessment of Overall Soil Corrosivity to Steel

1. pH	Points	4. Soil Type	Points
0-2	5	Clay (Blue-Gray)	10
2-4	3	Clay/Stone	5
4-8.5	0	Clay	3
8.5+	3	Silt	2
		Clean sand	0
2. Cl⁻, ppm		**5. Soil Resistivity**	
1,000+	10	<1000	10
500-1,000	6	1,000-1,500	8
200-500	4	1,500-2,500	6
50-200	2	2,500-5,000	4
0-50	0	5,000-10,000	2
		10,000+	0
3. Redox, mV (vs Cu:CuSO₄)		**6. Overall Rating**	
Negative	5	Severe	15+
0-50	4	Appreciable	10-15
50-100	3.5	Moderate	5-10
100+	0	Mild	0-5

20.3.1 Steels

Basically, steel items should not be exposed to soil without corrosion control measures as described below. This is because steel can be rapidly attacked under even temporarily aggravated conditions and because corrosion control is a very small part of the installed cost. Also, a buried structure or a tank bottom is going to be generally inaccessible for an extended period of time, so protective measures must be incorporated initially if a satisfactory life is to be obtained. Steels are also susceptible to corrosion by external influences (e.g., stray currents, telluric currents, galvanic effects, bacterial action, etc.). As described further in Chapter 38, steels are best protected underground by a combination of coatings and CP.

20.3.2 Cast Iron

Historical records indicate that cast iron pipe in the soil may last as little as 5 or as much as 75 years, depending upon specific conditions. Cast iron is inherently more resistant than steel, because of the more adherent nature of the rust formed under normal conditions. Nevertheless, it can suffer graphitic corrosion in soils of low pH or those saturated with soft aggressive waters. Corrosion can be aggravated by bacterial action. In modern usage, underground cast iron is conventionally protected by barrier coatings. CP is somewhat difficult because of the problem of establishing electrical continuity across the mechanical joints of cast iron pipe.

20.3.3 Zinc

Zinc may be quite resistant or poorly so, depending upon specific soil chemistry. From a practical standpoint, galvanized steel is not a good selection for underground service because the zinc coating is very thin and anodically active to every metallic structure in the vicinity. This does not preclude the use of galvanized steel structural legs for powerlines and the like, which can be further protected by coatings.

20.3.4 Aluminum

Aluminum gives useful service underground provided it is protected from galvanic demands and specific ion effects (e.g., chlorides). Great care must be taken in use of CP, because of the possibility of generating high alkali concentrations which cause corrosion of this amphoteric material.

20.3.5 Stainless Steels

All of the stainless steels are subject to pitting by high chlorides and/or oxygen concentration cells. Stray current effects are inherent in the use of DC reverse polarity welding of type 300 series stainless steel lines in the ground. Austenitic stainless steel piping should be coated and cathodically protected for underground installations. One never knows when corrosive conditions will arise, and the localized attack characteristic of conventional stainless steels can lead to rapid penetration.

20.3.6 Lead

Lead sheathing has been successfully used in telephone cable sheathing underground for many years. It is generally resistant except for stray current effects, but can be corroded by specific chemical contaminants.

20.3.7 Copper Alloys

The behavior of copper alloys in soil follows along the lines discussed for water. Soft, acidic waters arise in soil from organic matter deteriorating in marshlands. Sulfate-reducing bacteria (SRB) can cause corrosion by sulfides. If ammonia is formed by rotting of nitrogenous compounds, corrosion or SCC can ensue. Nevertheless, copper is usually satisfactory even in saline soils because of the static nature of the exposure, unless local cell action develops.

20.3.8 Concrete

The corrosion behavior of concrete in soil follows the principles described for water, especially as regards pH and sulfate effects. However, even sulfate-bearing soils are less aggressive than the corresponding waters under flowing conditions. Concrete cylinder pipe and cement-asbestos pipe have a long history of successful application in soil. It should be remembered, however, that the so-called cement bacillus (i.e., sulfate ion effects) was reported first in high-sulfate soils.

20.3.9 Plastics

Where conditions of temperature and pressure otherwise permit, plastic pipe and tanks (e.g., PE, PP, FRP) are ideal for service in soil. FRP underground tanks are largely replacing coated steel for gasoline storage. The major problem is of mechanical damage from rock fill and other foreign objects, due to ground subsidence or surface traffic imposing a loading force against the pipe or vessel wall.

20.4 Corrosion Control

All five of the standard corrosion control procedures apply, although some are more useful and prevalent than others.

20.4.1 Materials Selection

Most often, the engineer is faced with an established material, such as a steel pipeline, a stainless transfer line, a copper water or gas line.

Opportunities for materials selection are limited in soil-type applications.

20.4.2 Environmental Control

Usually, the soil conditions are relatively fixed. Occasionally, the corrosivity of the soil can be lessened by mixing extraneous materials in the back-fill, such as sand, lime, or water-repellents.

20.4.3 Barrier Coatings

Various types of barrier coatings, usually those with a high dielectric constant and good alkali resistance, are employed in soil service. These conjointly with CP (see below), are the major practical means of corrosion control. Their primary function is to decrease the current demand of the buried structure, facilitating CP, but they also serve to exclude specific corrosive species.

20.4.4 Electrochemical Techniques

CP, specifically, is the major means of corrosion control for underground structures, using either sacrificial anodes or an impressed current system (see Chapter 38).

20.4.5 Design

There is good and bad design of underground installations, as in other engineering works. Major aspects of good design, from the corrosion standpoint, include electrical grounding practices, allowance for thermal expansion and contraction, soil control and compaction, and the design elements inherent in the CP systems.

21

Atmospheric Corrosion

Atmospheric corrosion is the third area in which water plays a significant role, although the corrosion is also largely due to the 20% oxygen in air. The combined action of water and oxygen can be severe on the less resistant metals, and is exacerbated by certain contaminants. Atmospheric corrosion is unique in the sense that the corrosion products tend to remain on the metal surface, to a great extent. Therefore, a layer of such products become the rate-limiting mechanism for many metals upon exposure to the atmosphere. Atmospheric corrosion is a very large part of the overall cost of corrosion, affecting consumer items such as automobiles and appliances as well as industrial plants.

21.1 Types of Corrosion

All types of corrosion phenomena may be encountered, depending upon the particular materials and specific atmospheric contaminants involved. Because of the large amount of steel potentially subject to atmospheric corrosion, general corrosion is the rule. However, localized forms such as pitting, IGA and SCC may be encountered with susceptible alloys.

The possibility of galvanic corrosion is somewhat minimized because the electrolyte available consists only of a thin film of condensed or adsorbed moisture, instead of the freely conductive volume available under immersion conditions. There is also a possibility that the accumulation of corrosion products between the mating surfaces of dissimilar metals may spread them apart and break the electrical contact. This cannot be relied on, however, and galvanic corrosion must always be considered in design for atmospheric exposures.

21.2 Controlling Factors

Both chemical and physical factors affect behavior of materials in atmospheric exposure.

21.2.1 Chemical Factors

Oxygen

Oxygen is always available in atmospheric exposures, playing a dual role. As in immersion conditions, it functions as a cathodic depolarizer, aggravating the corrosion of iron and steel. On the positive side, it polarizes the anodic reaction of metals which form passive films, whence the good atmospheric corrosion resistance of aluminum, titanium, and the stainless steels. It can also influence the nature of complex salts, aiding in the protective nature of corrosion products on other nonferrous metals.

Water

Water (moisture) is the other overriding factor in the atmosphere. *Relative humidity* (RH), the amount of water vapor contained in the air relative to the saturation limit expressed as a percentage, is often the significant parameter. For carbon steel, there is a lower critical RH of approximately 65% (the RH of a saturated ferrous chloride solution) and a second at approximately 88% (the RH of a saturated ferrous sulfate solution). A marked increase in corrosion rate occurs at these values. Corrosion of steel is quite low below 50% humidity.

In many external situations, the RH may not be relevant because the metal is not at the temperature of the surrounding air. The metal may be quite warm and dry during the day, when the sun is shining, but cooler than the air at night due to radiant cooling, with dew or frost forming on it. This is the key mechanism. Of course, deposition of hygroscopic salts or contaminants will aggravate the situation by keeping the surface wet.

Although dry air is substantially noncorrosive, one should be aware that the metal surface can see *high* humidities in its vicinity (e.g., due to hygroscopic salts or adsorption in films) even though the humidity in the bulk atmosphere is low. This is a perennial problem in ballast compartments of barges or tanker ships, and also in large atmospheric storage of moisture on a metal surface cooled by thermal radiation. Even in the open desert, there can be condensation of moisture on a metal surface cooled by thermal radiation. Relative humidity may be high *locally*, as in the lee of a cooling tower, despite a dry prevailing wind.

Contaminants

Contaminants are inherent in atmospheric exposure. They may consist of volatile species or particulate matter and may be either of natural or synthetic origin. Those which are most commonly encountered are listed in Table 21.1.

Table 21.1

Atmospheric Contaminants

Type	Volatile	Particles
Natural	Carbon dioxide	Sea salts (NaCl)
	Sulfur dioxide	Ashes
	Hydrogen Sulfide	Dust
	Ammonia	
	Oxides of nitrogen	
Synthetic	Carbon dioxide	Coal dust
	Sulfur dioxide	Fly ash
	Sulfuric acid	Smoke
	Chlorine	Fumes
	Hydrogen chloride	Road salt
	Oxides of nitrogen	Cement dust
	Ozone	

Humans contaminate the atmosphere by their very presence, their flocks and herds contribute to the ammonia content, and fires to the oxides of carbon, sulfur, and nitrogen. Industry adds the more aggressive acidic species. Particulate matter such as elemental sulfur, coal dust, and fly ash (from combustion of coal) can *catalyze* certain corrosion reactions to promote atmospheric corrosion.

21.2.2 Physical Conditions

The degree of shelter from particulate fall-out; orientation relative to sunshine, rainfall, and prevailing wind; even the time of year of initial exposure can affect initial and ultimate corrosion rates.

21.3 Types of Atmospheres

Although meteorologists classify air systems as to temperature and moisture (e.g., polar, temperate, tropical, desert, marine), corrosion specialists have traditionally classified exposure conditions as *rural, marine, industrial,* and *indoor.*

A rural atmosphere traditionally was that of the inland farm. Marine atmospheres are associated with coastal areas of up to several miles inland, at least. Industrial atmospheres originally meant the coal-polluted atmospheres of commercial cities. An indoor exposure suggested the temperature, humidity, and freedom from contamination of an office or warehouse. Even in the original context, it is evident that there is no clear line of demarcation between these categories.

Thanks to increasing concern over the environment, the atmosphere is generally less polluted today. Nevertheless, there can be industrial pollution of rural atmospheres, which may also have routinely high humidities. Mixtures of marine and industrial effects are common in many areas, each type of contaminant aggravating the other. Whether marine atmospheres effectively cease 100 meters from the shore or ten kilometers inland depends upon the prevailing winds, their *fetch* (the distance traversed over water at speed), and the height of the local surf.

Indoor atmospheres may be as pristine as a controlled atmosphere room for precision watch-making or as severe as those around a pickling bath in a steel mill. Really, any atmospheric condition should be defined in terms of temperature, humidity, and contaminants *or* its corrosivity quantified in relation to the materials of interest (e.g., steel, zinc, aluminum, copper alloys). One should always consider special circumstances such as cooling tower drift or spray, or spills or releases of water or chemicals.

21.4 Corrosion Control

21.4.1 Materials Selection

This approach is used fairly often, although the materials of construction are often dictated by the end use of the artifact.

21.4.2 Environmental Control

Such an approach is used only in the case of air-conditioned buildings or special installations for finished machining operations.

21.4.3 Barrier Coatings

Paints and coatings are the major corrosion control technique employed against atmospheric corrosion.

21.4.4 Electrochemical Techniques

Cathodic protection in the conventional sense is not applicable to atmospheric corrosion because there is no bulk electrolyte to convey current. However, metallic coatings of an anodic nature relative to the substrate (e.g., zinc or cadmium on steel), as well as zinc-pigmented paint systems, will confer CP to the underlying metal at scratches, faults, or holidays in the coating.

21.4.5 Design

While certain metals and alloys have their own characteristic behavior in the atmosphere, proper design is important in many cases. For example, stainless steel or anodized aluminum in architectural applications can behave in a very disappointing manner unless the design permits regular washing or cleaning. Dirt or other films can cause pitting as well as unsightly stains. ASTM A 242[1] weathering steels must be boldly exposed (i.e., with no crevices) to develop their characteristic protective oxide film. Even then, water run-off can cause serious rust discoloration on adjoining concrete structures. Copper alloys cause a similar discoloration problem (green), but to a much lesser degree.

21.5 Specific Materials

Following is a brief resume of the important properties of specific metals and alloys in atmospheric exposure.

21.5.1 Magnesium

Despite its very anodic nature, magnesium will resist mild atmospheric exposure. It is used for ladders and other light-weight structural assemblies (e.g., in lighter-than-air craft).

21.5.2 Aluminum

Aluminum may often be freely exposed to the atmosphere, as with vessels, tank trucks, aircraft, etc. Performance will vary with the alloy and heat treatment (IGA and SCC may be problems in industrial or marine atmospheres) and with specific contaminants. Even in apparently similar marine atmospheres, there may be as much as a tenfold difference in pitting proclivity between different alloy and location combinations. Aluminum is amphoteric, being attacked by both acidic and alkaline contaminants. However, it is useful in hydrogen sulfide- and sulfur dioxide-type atmospheres. Chlorides are conducive to pitting and SCC, and aluminum is very subject to under-deposit or poultice attack. Anodizing, by chemical or electrochemical means, will reinforce the surface oxide film and improve performance in atmospheric exposure. Aluminum is also successfully used as a hot-dipped coating and metallic pigment for paints and coatings, as well as a metallized surface for some marine applications.

21.5.3 Iron and Steel

Except in the most innocuous atmospheres, iron and steel must be protected from corrosion by one of several means. Temporary rust preventatives, painting, galvanizing, and other protective coatings may be employed, depending upon the severity of the atmosphere and the required performance or life.

There is a group of low alloy steels of relatively high strength, known as *weathering steels* (ASTM standards A 242[1] and A 588[2]), which form self-protective rust films when boldly exposed to many industrial, as well as uncontaminated, atmospheres. An example is Corten[†]. Such steels are used for structural purposes and rolling stock, unpainted, where advantage can be taken both of their higher strength and improved corrosion resistance, as compared with carbon steel. They will not resist chloride-contaminated industrial or industrial-marine atmospheres. Crevices, as in riveted lapped joints, must be sealed with some sort of mastic against ingress of atmospheric moisture.

21.5.4 Zinc

Zinc has useful resistance in mild atmospheric exposures. It is used primarily as hot-dipped galvanizing and as a paint pigment, but electroplated zinc is also employed. In ordinary atmospheres,

[†] Trade name

corrosion is linear with time and the degree of protection of the steel substrate is a function of thickness of zinc, regardless of the method of application. As time passes, an accumulation of noble metal salts may cause a slight acceleration of attack.

The corrosion rate is quite low until the steel substrate is exposed, when galvanic effects take over, but corrosion will be exacerbated by acidic or alkaline species (zinc also being amphoteric). Galvanizing is inadequate for long-term marine atmosphere exposure.

21.5.5 Cadmium

Cadmium was at one time thought to be better than zinc, both in inherent resistance and as a sacrificial plating. This conclusion was drawn from salt spray tests, which are not confirmed in real-world atmospheric exposures. However, it is only available as an electroplate in thin films and is therefore inadequate except for mild indoor service. Because of the toxicity of cadmium, it is now rarely employed.

21.5.6 Lead

The only practical applications of lead are as terne-plate, a 4:1 lead-tin alloy used for roofing applications and to protect automotive brake-linings.

21.5.7 Copper

Copper and its alloys take on a protective patina in mild atmospheres, known as *verdigris* (French for "green-grey"). High-strength copper or copper alloys (e.g., silicon bronzes, yellow brasses) are susceptible to SCC due to trace amounts of ammonia and/or nitrites. (Chlorides seem to inhibit SCC somewhat, permitting use of 30% zinc brasses in marine atmospheres.) All acid fumes are corrosive, due to the synergistic effect of atmospheric oxygen and moisture. Hydrogen sulfide in the air will cause severe corrosion of copper under heavy black deposits, but yellow brasses may form a thin protective film.

21.5.8 Stainless Steels

Stainless steels will remain bright in uncontaminated atmospheres. Continued cleanliness is the key to successful performance. Even the austenitic grades, however, can rust and pit if chloride contamination is prevalent or if dirt or deposits are allowed to accumulate on the surface. Also, sensitized stainless steel will show SCC in an intergranular mode, while conventional chloride SCC has been observed in cold-work hardware (e.g., sailboat stays).

21.5.9 Higher Alloys

The high-nickel alloys (e.g., alloys 625 [UNS N06625], C-276 [N10276], C-22[†] [N06022], and 59 [N06059]) and titanium and its alloys should remain bright, retaining even a mirror finish in atmospheric exposure.

21.6   Special Problems

There are three special problems, arising from or akin to atmospheric corrosion, which should be mentioned.

21.6.1   Threaded Fasteners

Because of the stresses and crevices inherent in a nut-and-bolt assembly, corrosion of steel is greatly aggravated in aggressive atmospheres. Cadmium-plated steel is useful only for indoor and noncorrosive service. Galvanized steel bolts perform well in mild to moderate atmospheric service, but even cooling tower spray (let alone marine situations) quickly causes corrosion. In such fasteners, "freezing" of the nut causes problems, perhaps long before metal wastage itself is significant. Hot-dip aluminized bolts (e.g., Bethalume[†]) are superior in severe industrial/marine atmospheres. *Properly* coated bolts (e.g., polyimide/PTFE over cadmium-plated steel; Xylan[†]) are an acceptable and economical alternative in most cases.

21.6.2   Corrosion Under Thermal Insulation

Thermal insulation when exposed to water can hold a reservoir of available moisture on the metal surface which, together with the permeability of air, causes severe attack up to several *tenths* of an inch per year (200 to 300 mpy [4 to 6 mm/y]) particularly on warm steel surfaces. On cold insulated surfaces, severe corrosion may occur where structural members abut the insulated pipe or vessel, permitting rime ice to form. With other specific contaminants, insulation can cause SCC of high strength copper alloys and external SCC (ESCC) of type 300 series stainless steels. Aluminum is rapidly attacked in the presence of chlorides or alkaline contaminants. Vessels and large piping may be coated with an appropriate organic system (such as a catalyzed epoxy-phenolic and modified silicone), prior to insulating, to help prevent such corrosion in the event of ingress of water. For stainless steels, the coatings should be zinc- and chloride-free, as previously discussed.

21.6.3 Cryogenic Plants

Cryogenics relates to very low temperature operations, in which two types of problems are encountered. First of all, any structural members which are fastened to the vessels and extend out through the insulation are chilled by the service temperatures. Moisture tends to condense and freeze on the chilled surfaces. The immediately adjacent areas, as well as the frozen areas which thaw during shut-down, are subject to high rates of attack. High quality coating systems are required to withstand corrosion in near-freezing zones.

In some vessels, the top may be at sub-zero temperatures but the bottom at 40 to 80°C (105 to 175°F). In such cases, atmospheric moisture is drawn in through the insulation to freeze on the colder parts. Some of it melts (all of it during shutdown) and runs down behind the insulation to cause severe corrosion of steel. Steel and alloy-steel vessels in low temperature service should be coated with a good quality, heavy-duty paint system before the insulation is applied.

References

1. ASTM Standard A 242, "Specification for High-Strength Low-Alloy Structural Steel" (Philadelphia, PA: ASTM).
2. ASTM Standard A 588, "Specification for High-Strength Low-Alloy Structural Steel with 50 ksi (345 MPa) Minimum Yield Point to 4-in. Thick" (Philadelphia, PA: ASTM).

Suggested Resource Information

S. K. Coburn, ed., *Atmospheric Factors Affecting the Corrosion of Engineering Metals*, ASTM STP 646 (Philadelphia, PA: ASTM, 1978).

S. W. Dean, T. Lee, eds., *Degradation of Metals in the Atmosphere*, ASTM STP 965 (Philadelphia, PA: ASTM, 1988).

S. W. Dean, E. C. Rhea, eds., *Atmospheric Corrosion of Metals*, ASTM STP 767 (Philadelphia, PA: ASTM, 1982).

W. I. Pollock, J. M. Barnhardt, *Corrosion of Metals Under Thermal Insulation*, ASTM STP 880 (Philadelphia, PA: ASTM, 1985).

J. W. Slusser, et al., "Stress Corrosion Cracking of Copper Alloys and Nickel Alloys," CORROSION/86, paper No. 330 (Houston, TX: NACE International, 1986).

22

Oxidizing Acids

Because the corrosion of metals to their ions necessarily entails oxidation (i.e., the loss of electrons), there is some confusion over what constitutes corrosion under "oxidizing" conditions. Acids, after water, are the most common corrosives and it is customary to classify them as *oxidizing* or *reducing acids.*

With a reducing acid, while the anodic oxidation of the metal is occurring, the cathodic reaction is primarily the reduction of hydrogen ions to atomic, then molecular, hydrogen. This is exemplified by the evolution of hydrogen by zinc in dilute hydrochloric acid.

In an oxidizing acid, the cathodic reaction is the reduction of the acidic anion rather than hydrogen evolution. For example, brown oxides of nitrogen are liberated by the reaction of dilute to moderate concentrates of nitric acid on steel.

It is important to note, however, that the characteristics of a metal/acid reaction are also influenced by the nature of the metal component. For example, boiling 55% sulfuric acid is a reducing acid to steel or type 300 series stainless steel, liberating hydrogen, while it is oxidizing to the cast silicon-nickel alloy, the sulfate ion being reduced to sulfur dioxide and hydrogen sulfide (and elemental sulfur being formed as a consequence of their interaction). This is in accord with the generality than any "oxidizing agent" can be a reducing agent in the presence of a stronger oxidant.

In some combinations (e.g., type 300 series stainless steel vs dilute sulfuric acid), oxidizing cations like ferric or cupric ions move the redox potential in an oxidizing direction by providing an alternative cathodic reaction (i.e., reduction of cupric or ferric ions) to hydrogen evolution. We then have oxidizing conditions although not, strictly speaking, an oxidizing acid.

Solutions of oxidizing acid salts (e.g., ammonium nitrate) act like dilute solutions of the parent acid.

In general, oxidizing acids tend to corrode metals which do not form a passive oxide film (e.g., copper, lead) rather than those which do, such as chromium-bearing alloys, titanium, and aluminum. Reducing acids are sometimes *more* aggressive to the normally passive metals than to active metals, because of the reaction of nascent hydrogen with the oxide film or direct hydriding of the metals itself, as with titanium, zirconium, and tantalum. Some of the more important oxidizing acids are discussed in detail below.

22.1   Nitric Acid

Nitric acid is not only a strong mineral acid but a powerful oxidizing agent, even in dilute solutions. Most nitric acid is manufactured by a process involving oxidation of ammonia to give a product of approximately 60% concentration. This is purified and concentrated to give reagent grade, chemically pure (C.P.) acid of 70% concentration. Very strong acid in the 90 to 100% range can be made by dehydrating weaker solutions with concentrated sulfuric acid. Acid above approximately 85% is known as *fuming nitric acid* because it gives off red or white oxides of nitrogen (e.g., nitrogen tetraoxide). The corrosion characteristics of very strong nitric acid are somewhat different than those of more dilute concentrations because of an excess of nitronium ions (NO_2^+) over hydronium ions (H_3O^+).

22.1.1 Materials of Construction

The following discussion relates to the corrosivity of pure nitric acid to the common materials of construction, except where otherwise noted. Contaminants (e.g., halogens, halides, or oxidizing cations) can profoundly alter the expected corrosion behavior.

Aluminum

Aluminum and its alloys have good resistance to >80% acid at room temperature and to fuming acids in the 93 to 96% range to approximately 43°C (110°F). Above 96%, even higher temperatures may be tolerated. The alloys most frequently used are A91100, A93003, A95052, and A95454. It is important to employ low-silicon welding rods (e.g., A91100, A95386) and not to permit localized dilution, as by leakage or by ingress of moist air, as this will cause very rapid attack.

Iron and Steel

Iron and steel, even when "passivated" by cold concentrated nitric acid, with no evolution of brown fumes, show corrosion rates too high for practical consideration, in most instances. Very strong "mixed acid," containing less than 2% water and approximately 15% sulfuric acid has been handled in steel equipment. However, vapor-phase corrosion may occur. Dilute acid attacks cast iron and steel very rapidly. The relative activity of intermediate concentrations of acid can be used to differentiate low alloy chromium steels (1 to 9%) by spot test; freshly abraded 12% chromium stainless steel is unaffected by 20% acid, being passivated rather than corroded.

Silicon Cast Irons

Silicon cast irons (such as 14% silicon, [UNS F47003]) have outstanding resistance to acids above 45% to the atmospheric boiling point. The resistance *increases* with acid concentration, the rate being substantially nil in strong acid at high temperatures. The corrosion resistance is due to formation of an adherent siliceous film.

Stainless Steels

Stainless steels of the ferritic type (e.g., type 430 17%-chromium stainless steel, [S43000]) were among the first used in nitric acid service. Welding problems, and the brittleness of the cast forms, ultimately led to their replacement with the austenitic grades.

The molybdenum-free austenitic stainless steels of the type 300 series variety are outstanding for their resistance to nitric acid in the annealed condition. However, the potential for IGA is high, and the low carbon type 304L or stabilized grades (e.g., type 347) are required for welded equipment which is not amenable to solution annealing. A nitric acid grade (NAG), which has <0.02 C and very low phosphorus by virtue of AOD production, is currently used.

IGA can also occur, regardless of composition or heat treatment, if hexavalent chromium (CrVI) ions accumulate in the acid to some concentration level. Wrought type 316L should not be used in nitric acid because of corrosion of sigma phase, although annealed castings (CF-3M) are acceptable if properly heat-treated.

Higher alloy grades, such as type 310L have been used at elevated temperatures; a NAG variant is available. New developments in silicon-rich stainless steels (18%Cr-18%Ni-5%Si) are replacing the 25%Cr-20%Ni alloys above 70% nitric acid at>80°C (175°F).

Stainless steel tanks have been severely attacked in the *vapor* phase of strong nitric acid, 93 to 99%, apparently due to an autocatalytic effect of the nitric oxides.

Titanium

Titanium is highly resistant to nitric acid below 25% and in the 65 to 90% range at the atmospheric boiling point. In 25 to 50% boiling acid, rates are 0.25 mm/yr (10 mpy) or more. Titanium also resists fuming acid but there is danger of violent pyrophoric reactions if the water content is less than approximately 1.3% or the nitrogen dioxide content is greater than 6%. SCC can also occur in red fuming nitric acid. Only a professional corrosion engineer should select materials for this type of service.

Other Metals

Copper and nickel alloys (except for the chromium-bearing varieties, such as alloys 600 (UNS N06600), 625 (N06625), and C-276 (N10276), are rapidly attacked by even dilute nitric acid. The chromium-bearing nickel alloys are not usually economically attractive compared with stainless steels.

Lead is nonresistant in nitric acid.

Of the reactive metals, zirconium is better than titanium in 65 to 90% acid. However, it is subject to SCC at concentrations greater than 70% and fails rapidly in boiling 94% nitric acid. Tantalum has excellent resistance up to the atmospheric boiling point.

Of the noble metals, gold and platinum are resistant but silver is rapidly attacked. A 3:1 mixture of hydrochloric and nitric acid, aqua regia, is the classical solution for dissolving gold, due to oxidation of the HCl to nascent chlorine.

Nonmetallic Materials

PTFE, plain or glass-filled, is routinely employed in nitric acid service. Carbon is a useful material, provided it is free of oxidizable binders. Other organic materials are limited by the temperature and concentration of the acid, which control its oxidizing capacity. Some suggested limitations are given in Table 22.1.

Table 22.1

Nonmetallics in Nitric Acid Service

Material	Concentration at Ambient Temperature 25°C (77°F)	Concentration at Elevated Temperature
PTFE	100%	100% @ 260°C
FEP	100%	100% @ 200°C
PVC (unplast.)	50%	40% @ 60°C
PE or PP	60%	20% @ 40°C
Butyl rubber	50%	30% @ 60°C
Karbate[†]	30%	10% @ 85°C

22.1.2 Handling and Storage

Following is a listing of materials of construction which are thought to constitute good engineering practice for a variety of items, subject to specific limitations previously discussed:

Tanks	304L; aluminum (over 93%);
Piping	304L;
Valves	CF-3 (J92700); CF-3M (J92800) acceptable;
Pumps	CF-3 or CF-3M; titanium, and
Gaskets	Spiral-wound stainless steel with PTFE.

22.2 Chromic Acid

Chromic acid is a powerful oxidizing acid but requires more discrimination than does nitric acid in materials selection. In general, materials attacked by oxidizing acids (e.g., copper, nickel) are unsuitable.

Aluminum

Aluminum may be used to approximately 10% concentration to approximately 66°C (150°F).

[†] Trade name

Iron and Steel

Conventional steels and cast irons are attacked by chromic acid solutions but the 14% silicon cast iron (UNS F47003) has been widely used to approximately 40°C (104°F).

Stainless Steels

At room temperature, the molybdenum-free type 300 series grades may be used to approximately 30% concentration, but rates increase dramatically above 5% and 80°C (175°F).

Reactive Metals

The common reactive and refractory metals (Ti, Zr, and Ta) are resistant up to 50% chromic acid to approximately 100°C (212°F).

Precious metals

Silver will resist all concentrations of chromic acid to the atmospheric boiling point.

Other Metals and Alloys

Strangely, magnesium is highly resistant in the absence of chloride ion contamination. A boiling 20% solution has been used to clean magnesium alloys without attacking the base metal.

Tin will resist chromic acid to approximately 80% and 100°C (212°F), and a lead alloy containing 7% tin has been used to handle chromic acid solutions. Lead itself will resist chromic acid to 85% at 220°C (430°F), 93% at 150°C (300°F), and 95% at ambient temperatures. Corrosion rates increase, although still acceptable, in concentrations below approximately 5%.

Nonmetallic Materials

Conventional plastics (e.g., CPVC, PE, PP) are limited to not more than 50% acid at or below approximately 70°C (160°F), but fluorinated plastics are acceptable to their normal temperature limits.

Glass and ceramics are fully resistant, although a special iron-free chemical stoneware tile is specified for plating solutions.

22.3 Concentrated Sulfuric Acid and Oleum

One cannot set the limits for oxidizing characteristics of sulfuric acid independently of the materials to which it is exposed. However, it *starts* to have a definite oxidizing nature at approximately 5 Normal (25%), being reduced by nickel or alloy 400 at the boiling point and at room temperature by finely divided ("Raney") nickel. By 60%, at approximately 80°C (175°F), it will carbonize PVDC over a prolonged exposure period. At 95% and 25°C (77°F), it carbonizes FRP instantaneously. Concentrated sulfuric acid can be considered 70 to 100% acid. Oleum is 100% acid plus dissolved sulfur trioxide, and is generally designated as, for example, 20% oleum, which is equivalent to 104.5% acid if all the SO_3 is converted to H_2SO_4.

Dilute acid (e.g., below 25%) and the intermediate strengths between 25 and 70% will be considered in the Chapter 23 as *Reducing Acids*, even though specific oxidizing ions or other contaminants can radically alter the corrosion characteristics.

22.3.1 Materials of Construction

The behavior of common materials in concentrated sulfuric acid and oleum is discussed briefly below.

Aluminum

Aluminum and other light metals are, for practical purposes, nonresistant to concentrated acid below 96%. It has been used successfully above 96%, even at elevated temperature, provided there is no opportunity for inadvertent dilution.

Cast Iron and Steel

Cast iron and steel are just beginning to be useful at 70% concentration. Steel is more resistant in the 70 to 80% range than in 80 to 90% acid. Concentrations above 90% are routinely handled in iron and steel *provided* velocities are below approximately 0.7 m/s (2 ft/s) for steel and below 1.5 m/s (5 ft/s) for ductile cast iron, whose ferric sulfate film is more tenacious. (Conventional gray cast iron is no longer used in modern applications, for safety reasons.) Steel shows a decided corrosion peak at approximately 101% sulfuric (5% oleum) but can be used in 20 to 30% oleum.

Ordinary gray cast iron is totally unacceptable in oleum because of internal corrosion by the free sulfur trioxide along the graphite flakes. The gaseous products plus iron sulfates and silica (from oxidation of silicon) have caused explosions of conventional cast iron vessels. For this reason, ductile cast iron or a specialty material, Procon[†], *must* be used in oleum.

In storage of concentrated acid, localized dilution by ingress of atmospheric moisture can cause attack in the vapor space under high-humidity conditions, unless a *dessicating* vent is provided.

Silicon iron (14.5% [UNS F47003]) is resistant up to the boiling point in concentrated sulfuric acid, and would replace ordinary iron or steel at temperatures above approximately 50°C (122°F). It is, however, a very brittle material and must be protected from thermal or mechanical shock. The high-silicon iron must not be exposed to oleum.

Austenitic nickel cast irons, such as UNS F41000 and F41004, are successfully used to resist corrosion, erosion, and abrasion in oleum at ambient temperature.

Lead

Lead has been a longtime favorite in the sulfuric acid industry, only recently falling somewhat into disfavor because of toxicity during joining ("lead-burning"). Depending on a sulfate film for resistance, like steel, lead is very susceptible to velocity effects above approximately 1 m/s (3.3 ft/s). Also, solubility of lead sulfate increases sharply at and above 95% concentration and also at lower concentrations (e.g., 80%) at approximately 120°C (250°F). It is not resistant to oleum.

Austenitic Type 300 Series Stainless Steels

Austenitic type 300 series stainless steels are resistant to cold concentrated sulfuric acid. Type 304L is routinely used for piping in sizes below 75 mm (3 in.), where cast iron is not available. With increased velocity and turbulence, higher alloys are employed (e.g., CF-3M for valves, CN-7M [N08007] for pumps). The molybdenum-free grades (i.e., types 304 and 304L) should be used in concentrations greater than 93% acid and in oleum. In oleum of less than 14% concentration, corrosion may be aggravated by velocity effects at temperatures as low as 60°C (140°F). Higher alloys like type 309 (S30900) are successfully used at elevated temperatures and the new 5% silicon grades resist sulfuric acid better than conventional grades.

Copper and Nickel

Copper alloys are not used in concentrated acid, because of its oxidizing nature.

The lower nickel alloys are nonresistant to oxidizing acids and the chromium-bearing nickel alloys are not economical in conventional applications. However, alloy B-2 (UNS N10665) is resistant to the atmospheric boiling point unless traces of nitric acid or ferric ion contamination are present. It is not resistant to oleum, however. Nickel-chromium-molybdenum alloys (e.g., alloy C-276 [UNS N10276]) will withstand oleum, but they are not usually economically competitive with lower-cost materials.

Reactive Metals

Tantalum will withstand 95% acid to 175°C (350°F) and lower concentrations to the atmospheric boiling point. It is attacked by sulfur trioxide, making it unsuitable for oleum service, and suffers hydrogen attack in galvanic couples. Titanium and zirconium are not resistant in concentrated acid, as defined.

Noble Metals

Gold (UNS P00010) and platinum (P04898) have been widely used in sulfuric acid concentrators, but silver is not resistant.

Nonmetallic Materials

Pure carbon is resistant to boiling 100% acid and resists 115% acid to approximately 70°C (160°F). Impervious graphite (e.g., Karbate†) will perform up to approximately 150°C (300°F) *unless* it has cemented joints, which impose a limit of 60°C (140°F).

Glass, porcelain, stoneware, and acid-brick are useful in many applications involving hot concentrated acid.

Fluorinated plastics (e.g., PFA, PTFE, FEP, PVDF, Kynar†) in acid are restricted only by their inherent temperature limitations but permeability poses problems in oleum service.

Other plastics (e.g., PE, PP, PVDC, PVC) will withstand 75% acid to 50°C (120°F) and 90% acid at 30°C (85°F); above these limits, carbonization will occur. FRP, with a suitable resin, will withstand 75% acid and 25°C (77°F) maximum. They will not withstand oleum.

Elastomers, other than the fluorinated variety, are limited to a maximum of 75% acid and not more than 80°C (175°F) even at 70%. However, Kalrez[†] 1045, a fully fluorinated elastomer filled with titania, can be used for oleum service.

22.3.2 Handling and Storage

Following is a list of materials of construction for various items of equipment, which list is considered good engineering practice for the handling and storage of concentrated acid at ambient temperatures:

Tanks	Carbon steel (with drying vent); coat with baked phenolic or anodically protect (AP) if iron contamination is objectionable;
Piping	Ductile cast iron (304L for small diameter);
	Note: electric or hot-water tracing only.
Valves	CF-3M (or CN–7M for "throttling");
Pumps	CN-7M;
Gaskets	Stainless steel/spiral-wound PTFE; and
Dilution	PTFE-lined or Kynar[†]-lined; alloy C-276 tee check valves.

Similar recommendations for oleum are as follows:

Tanks	Carbon steel;
Piping	304; velocity max 1.3 m/s (4 ft/s);
Valves	CF-8M;
Pumps	CF3M or Pro-Iron[†] with Lewmet[†] impellers and 20Cb-3[†] (alloy 20) shafts, and
Gaskets	Kalrez[†] 1045; spiral-wound 304/ PTFE

Suggested Resource Information

ASM Handbook, Vol. 13, *Corrosion* (Materials Park, OH: ASM International, 1987).

C. P. Dillon, *Materials Selection for the Chemical Process Industries* (New York, NY: McGraw-Hill, 1992).

M. G. Fontana, *Corrosion Engineering,* 3rd Ed. (New York, NY: McGraw-Hill, 1986).

B. J. Moniz, W. I. Pollock, eds., *Process Industries Corrosion—The Theory and Practice* (Houston, TX: NACE International, 1986), pp. 243 and 259.

NACE Technical Committee Report 5A151, "Materials of Construction for Handling Sulfuric Acid" (Houston, TX: NACE International)

L. S. VanDelinder, *Corrosion Basics–An Introduction* (Houston, TX: NACE International, 1984).

H. H. Uhlig, R. W. Revie, *Corrosion and Corrosion Control,* 3rd Ed. (New York, NY: Wiley-Interscience, 1985).

Recommended Software Program

CHEM•COR[†] 1 and CHEM•COR[†] 1 PLUS, Concentrated Sulfuric acid and Oleum (Houston, TX: NACE International).

23

Reducing Acids

The non-oxidizing or reducing acids are the inorganic and organic acids which characteristically evolve gaseous hydrogen during the corrosion of active metals. They are corrosive to metals above hydrogen in the electromotive series only in the presence of oxygen or oxidizing agents, whose reduction substitutes for hydrogen evolution. The behavior of passive metals and alloys may be fairly unpredictable, depending on acid concentration, temperature, DO, and specific contaminants. (*Note:* The acid *gases*, carbon dioxide and hydrogen sulfide, are covered in separate chapters, although their water solutions fall into this category of reducing acids.)

23.1 Inorganic Acids

Inorganic acids include the mineral acids (i.e., low concentrations of sulfuric acid, phosphoric acid) and hydrochloric acid. Other reducing acids will have similar corrosive properties.

23.1.2 Hydrochloric Acid

Hydrochloric acid is an aqueous solution of hydrogen chloride. As the concentrated acid (36%), it is a pungent liquid. The "constant boiling mixture" (CBM) is 22% at atmospheric pressure; concentrations above 22% give off hydrogen chloride to reach the CBM. Dilute solutions tend to evaporate water to reach the same value.

A highly corrosive acid in its own right, the corrosion behavior can be drastically altered by contaminants. Muriatic acid is a commercial 30% acid, heavily contaminated with dissolved ferric iron (Fe III) salts. Many "by-product" acids are heavily contaminated. Trace amounts of chlorinated solvents or aromatic solvents profoundly influence the resistance of plastics and elastomers to what is nominally hydrochloric acid. The discussion below refers specifically to pure hydrochloric acid, except where otherwise noted.

Specific Materials

The behavior of specific categories of materials in hydrochloric acid is as follows:

(1) *Light Alloys.* Aluminum and magnesium alloys are severely attacked by hydrochloric acid.

(2) *Iron and Steel.* These metals are inherently nonresistant to hydrochloric acid. However, steel piping and vessels can be chemically cleaned with *inhibited* acid for a few hours at a time up to 65°C (150°F) under controlled conditions (e.g., controlled velocity; no cast iron or stainless components in the system). Inhibitors are ineffective with cast iron, because of the galvanic influence of the contained graphite.

(3) *Stainless Steels.* Stainless steels of all kinds are incompatible with hydrochloric acid. Further, residual chlorides can contribute to pitting or SCC in subsequent service (e.g., after acid cleaning).

(4) *Copper Alloys.* Copper and its alloys are attacked by hydrochloric acid in the presence of DO or oxidizing ions. Since the cupric ion is itself an oxidant, few practical applications will be found. Dealloying is also a potential problem (e.g.,dezincification, dealuminumification, destannification) for brasses and bronzes.

(5) *Lead.* Lead shows reasonable resistance in laboratory tests up to approximately 30% acid at 25°C (77°F) and to 20% at 100°C (212°F) but field experience has not been good. Corrosion products are quite soluble and easily washed away by flow.

(6) *Nickel.* Nickel and its alloys are superior to copper but not really useful until the 30% molybdenum alloy B-2 (UNS N10665) is employed (although alloy 600 [N06600] may replace type 300 series stainless steels where only trace amounts of acid are encountered). Alloy B-2 (N10665) will resist boiling hydrochloric acid, but *not* if traces of oxidizing agents (e.g., Fe III) are present. The chromium-bearing grades (e.g., alloy C-276 [N10276], C-4 [N06455]) will resist dilute acid plus ferric chloride, but only to intermediate temperatures.

(7) *Reactive Metals.* Titanium is non-resistant, but zirconium can withstand concentrated acid to 107°C (225°F) provided there are no more than 50 ppm oxidizing species (e.g., Fe III, Cu II) present. With oxidizing species present, *pyrophoric* corrosion products may be formed. In the absence of oxidants, zirconium will withstand 37% acid to 120°C (250°F), 25% to 160°C (320°F) and 15% acid to 200°C (390°F). However, unless high-purity material is used, a sort of "weld decay" may be encountered in the heat-affected zone (HAZ). Tantalum offers useful resistance to approximately 175°C (345°F), but is attacked by HCl *vapors* as low as 130°C (265°F).

(8) *Noble Metals.* Platinum will resist concentrated acid to 300°C (570°F). However, silver and gold will withstand only room-temperature service.

(9) *Nonmetallic Materials.* Glass and other ceramic materials are very resistant. External spillage or vapors corroding glass-lined steel equipment will generate nascent hydrogen, which penetrates the steel and dimerizes at the internal interface to cause internal spalling of the glass coating.

Rubber-lined equipment is traditional for handling acid up to 80°C (175°F). However, organic solvent contaminants (e.g., chlorinated hydrocarbons or aromatic solvents such as benzene or toluene) can be preferentially absorbed and concentrated to cause failure either of the rubber or of its adhesive.

Plastics (e.g., PVC, PE, PP) are resistant and FRP tanks and piping (either custom-built or fabricated with a PVC inner layer) are routinely employed to handle concentrated HCl. Note that standard commercial tankage, such as is used for water, gasoline, etc. is not suitable.

Handling and Storage

Following is a listing of materials of construction for various items, which are thought to constitute good engineering practice with minimum risk:

Tanks	FRP, rubber-lined steel;
Piping	FRP, polypropylene-lined steel;
Valves	Alloy B-2 (N-12M casting), PTFE-lined or FEP-lined;
Pumps	Alloy B-2 or N-12M, impervious graphite, PTFE-lined; and
Gaskets	Rubber, felted PTFE, FEP envelope, flexible graphite.

23.1.2 Hydrofluoric Acid

Both anhydrous hydrogen fluoride (AHF) and its 70% HF aqueous solution are commercially available. The laboratory grade is 48% and is also used in some industrial applications. The largest usage has been as an alkylation catalyst for gasoline and in the manufacture of chlorfluorocarbon refrigerants and propellants. (*Note:* Hydrofluoric acid and hydrogen fluoride are extremely toxic and very dangerous to personnel, producing painful and slow-healing burns.)

Specific Materials

The behavior of specific categories of materials in hydrofluoric acid is as follows:

(1) *Light Metals*. Magnesium, despite its anodic nature as a general rule, will resist hydrofluoric acid up to approximately 2% concentration, due to a film of insoluble corrosion products. This is of academic interest only.

Aluminum and its alloys should not be exposed even to dilute concentrations of hydrofluoric acid.

(2) *Iron and Steel*. Steels are resistant to concentrated acid (e.g., > 64% minimum) up to approximately 32°C (90°F). The resistance is due to a protective film of corrosion products, so velocity conditions cannot be tolerated. Hydrogen blistering may occur and welds may be preferentially attacked. *Hardened* steels are susceptible to environmental cracking (HAC).

Cast irons and alloy irons should not be used in hydrofluoric acid service, because of both corrosion and safety considerations. The 14% silicon irons (UNS F47003) are non-resistant, because of dissolution of the siliceous film.

(3) *Lead*. Lead was for many years the conventional material for handling hydrofluoric acid solutions, resisting up to 60% at 25°C (77°F). Corrosion rates are acceptable up to 25% acid and 80°C (175°F). Attack increases with acid strength, temperature and velocity. Anhydrous hydrogen fluoride rapidly attacks lead.

(4) *Copper*. Copper and its alloys are corroded to the extent that oxygen or oxidants are contained in the acid. Although not usually considered for this type of service, except as

flexible tubing and for AHF distillation, protective surface films may permit their use in certain processes containing HF as a reactant.

(5) *Stainless Steels.* Stainless steels of all types are unreliable, martensitic grades subject to HAC and others subject to pitting and/or SCC as well as general corrosion. However, type 304L resists AHF to 100°C (212°F) and the equivalent cast alloys (UNS J92900) have been used for pumps in this service.

(6) *Nickel Alloys.* Nickel 200 (N02200) will resist anhydrous HF up to approximately 150°C (300°F), but its usefulness in aqueous solutions is limited to non-oxidizing condition below approximately 80°C (175°F). It is both less resistant and more expensive than alloy 400.

Alloy 400 (N04400) has long been used for all concentrations to temperatures up to 120°C (250°F), although it is subject to SCC in the vapors in the presence of air, due to formation of cupric fluoride.

Alloy 600 (N06600) resists dilute aqueous solutions and AHF, in which it is used as valves to avoid the SCC problem with alloy 400, but is unreliable in intermediate concentrations due to its pitting propensities. Although used commercially in hot gaseous HF, it is not otherwise an economical choice.

Alloy B-2 (N10665) will resist the HF-sulfuric acid solutions encountered in the manufacturing process. Other nickel-chromium-molybdenum alloys (e.g., alloys C-276 and 625) are neither useful nor economically competitive with the lower nickel alloys.

(7) *Reactive Metals.* Titanium, zirconium, and tantalum are all severely attacked by even traces of fluorides, with hydriding and embrittlement.

(8) *Precious Metals.* Traditionally, silver, gold, and platinum have been used to handle hydrofluoric acid. "Fine silver" is recommended for pure HF, rather than "sterling silver" with its small amount of copper, but this is not true in some process mixtures.

(9) *Nonmetallic Materials.* Plastics without *hydroxyl groups* are very resistant and polyethylene has replaced the traditional wax-lined glass as the laboratory container for concentrated HF. Polystyrene, methacrylates, and vinyls can be used for up to 60% acid to 50°C (120°F). Phenol-formaldehyde plastics may be used up to 130°C (265°F), while fluorinated plastics are resistant up to their temperature limits. Because HF has a tremendous appetite for water, it rapidly attacks hydroxylated materials like polyesters. The fluorinated plastics are resistant but subject to permeation.

Natural rubber is used to ship 48% acid, and hard rubber is better than soft Neoprene in this service. The synthetic soft rubbers, such as butyl and neoprene, will withstand 60% acid to 70°C (160°F), and chlorobutyl rubber is used to ship 70% HF, although there may be vapor-phase attack. The compounding is critical; silica and magnesia must not be used. Despite its fluorinated structure, conventional Viton[†] is severely attacked by AHF at room temperature. There is a special grade available to resist AHF at room temperature. The perfluorelastomer, Kalrez[†], will resist AHF to 100°C (212°F).

Glass (and other siliceous ceramics) is rapidly attacked by HF, as well as by fluoride contaminants in other acids.

[†] Trade name

Carbon and graphite are resistant, but the *impregnated* impervious graphite (e.g., Karbate[†]) is limited to boiling 48% acid or to 60% acid at 80°C (185°F).

Wood is charred almost instantaneously by anhydrous HF.

Storage and Handling

Following is a listing of materials of construction for various types of equipment at ambient temperatures, which listing is considered good engineering practice with minimum risk:

Tanks	Steel[(A)];
Piping	Alloy 400 or PTFE-lined steel; steel[(B)] for AHF (max. velocity 1.5 m/s);
Valves	PTFE-lined, alloy 400 or steel with resistant trim (alloy 400 or alloy 20Cb-3[†] [alloy 20]);
Pumps	Alloy 400; and
Gaskets	Flexible graphite or spiral-wound alloy 400/PTFE.

[(A)] Steel tanks must be made with inclusion shape control (calcium-treated, argon-blown in ladle) with low sulfur and carbon equivalents. Fine grain size is required, and the vessels are fabricated from normalized and thermally stress-relieved.

[(B)] Piping can be carbon steel for AHF 1.5 m/s max.

23.1.3 Phosphoric Acid

Phosphoric acid is a syrupy liquid whose process of manufacture profoundly affects its corrosion characteristics. Acid made by the "wet" process of digestion of phosphate rock with sulfuric acid is heavily contaminated with impurities such as fluorides, chlorides, sulfates, and metal ions. Used mostly in the fertilizer industry, its corrosion characteristics are highly variable. Acid made directly from combustion of phosphorus is much more predictable, although it too can at times be a problem as to predictability.

Specific Materials

Behavior of specific categories of materials in phosphoric acid is as follows:

(1) *Light Metals.* Aluminum and magnesium are of no practical interest in phosphoric acid services, although aluminum can resist up to 20% acid to 65°C (150°F).

(2) *Iron and Steel.* Steel forms a protective film in acid above 70%, but is not used for this service because of iron contamination. However, a 1% solution is sometimes used as a "wash-coat" to prepare a steel surface for painting, and cold syrupy phosphoric acid (e.g., 85%) will *phosphatize* a steel or iron surface against rusting (if subsequently oiled) in indoor atmospheres for a period of time.

Cast iron is similar to steel in phosphoric acid service. The austenitic nickel cast irons, such as UNS F43000, are resistant to slightly above room temperature. The 14% silicon iron (F47003) will resist all concentrations to the atmospheric boiling point, but this resistance is contingent upon no *fluoride* contamination.

(3) *Stainless Steels.* Of the stainless steels, the 12% and 17% chromium grades have poor resistance to phosphoric acid. Despite optimistic laboratory data, field experience with type 304 or 304L has also been poor. Type 316L is usually reliable for tanks, piping, valves, and pumps handing *uncontaminated* acid up to 85% to approximately 80°C (175°F) maximum, although erosion-corrosion may occur at velocities in excess of 1 m/s (3.3 ft/s). Type 317L (S31703) or a high-performance grade (e.g., 20Cb-3[†][N08020]) may be substituted for an extra margin of resistance. The *molybdenum-bearing* high alloy grades (e.g.,

alloys 825 [N08825] and 20Cb-3†) will resist up to 85% acid at the atmospheric boiling point (in the absence of aggressive contaminants).

(4) *Lead.* Lead is a traditional material for handling phosphoric acid (particularly prior to the development of modern high nickel alloys) and will withstand 80% pure (or 85% impure) acid up to approximately 200°C (390°F). However, the resistance is due to films of insoluble lead phosphates, so erosion or impingement effects will cause problems.

(5) *Copper.* Copper and its alloys are governed in their corrosion behavior entirely by the influence of dissolved oxygen or other oxidants. Alloys immune to dealloying phenomena are useful in all strengths of acid to approximately 80°C (175°F) in the absence of oxidizing species.

(6) *Nickel.* Nickel (N02200) has very limited application in phosphoric acid, but alloy 400 (N04400) will withstand all concentrations to approximately 90°C (200°F) *if* there are no stronger oxidants present than dissolved oxygen and the cupric ion corrosion products do not accumulate. Alloy B-2 (N10665) is resistant to all concentrations to approximately 65°C (150°F) and up to 50% acid to the atmospheric boiling point, but only in the absence of strong oxidants (e.g., Fe III, Cu II). Alloy 600 (N06600) is not very useful, but its molybdenum-bearing variant, alloy 625 (N06625), has been successfully used in wet-process evaporators. Alloy G-30† (N06030) is often the best and most economical alloy for wet-process acid. The higher Ni-Cr-Mo alloys (alloy 625 [N06625], C-276 [N10276], etc.) are seldom used.

(7) *Reactive Metals.* Among the reactive metals, titanium is non-resistant (unless protected by

oxidizing contaminants). Zirconium is useful up to approximately 60% concentration, in the absence of fluorides. Tantalum will withstand any concentration up to approximately 175°C (345°F) but *only* if the fluoride concentration (a common contaminant) is less than 10 ppm. Otherwise, hydrogen embrittlement and pitting will occur.

(8) *Noble Metals.* Noble metals—silver, gold, and platinum—will resist all concentrations at least up to the atmospheric boiling point.

(9) *Nonmetallics.* Glass in its several applications (e.g., Pyrex[†], glass-lined steel) may be used to handle fluoride-free acid up to 60% and 100°C (212°F). At higher concentrations or temperatures, increasing rates of attack will occur.

Plastics and elastomers, as well as carbon and graphite, are useful within certain temperature and pressure limits.

FRP tanks and piping are widely used for atmospheric storage. Carbon and graphite are usually resistant to at least 350°C (700°F), but tube-and-shell heat exchangers may be limited by cemented joints.

Some recommended temperature limits are as follows:

Table 23.1

Materials in HF Environments

Material	Temperature Limits	
	°C	°F
PTFE	260	500
FEP	205	401
Polyesters	95	203
PE, PP, PVC	60	140
Elastomers	60	140

Handling and Storage

Following is a listing of suggested materials of construction thought to constitute good engineering practice for a variety of items for uncontaminated acid, with minimum risk:

Tanks　　FRP or 304L;

Piping　　FRP or 316L;

Valves　　CF-3M;

Pumps　　CF-3M; and

Gaskets　Elastomeric or graphic fiber.

23.1.4 Sulfuric Acid

Dilute and intermediate concentrations (i.e., less than 70%) of sulfuric acid are entirely different from the oxidizing concentrated acid discussed in the previous chapter. The most important thing to recognize is that its corrosive nature is affected *both* by dilution (and temperature) and by contaminants. The latter profoundly affect the redox potential of the solution, and we will discuss in this section both oxidizing and reducing contaminants, while maintaining that dilute sulfuric acid is usually in itself a reducing acid, particularly in relation to ferrous alloys.

It should first be understood that there is a relationship between specific materials and various concentrations of sulfuric acid as to whether hydrogen gas is evolved (as from a reducing acid) or whether the sulfate ion is reduced. (The behavior of specific metals and alloys is discussed under "Specific Materials" on the next page.) For example, some authorities indicate that sulfuric acid *starts* to become an oxidizing acid at and above 5N concentration (approximately 25%). This may be based on measured "redox potential"[1] or on some thermodynamic calculation. Certainly, a broad range of concentrations evolve hydrogen during the corrosion of ferrous alloys. It is evident, however, that finely divided *nickel* (e.g., "Raney" nickel) will react with 25% acid at room temperature, liberating hydrogen sulfide by reduction of the sulfate ion. At the boiling point, 25% acid undergoes anion reduction in the presence of alloy 400 (N04400). Boiling 53 to 57% acid is similarly reduced during corrosion of the now obsolete high-silicon cast nickel alloy, Hastelloy[†] D.

Over the entire range of dilution from approximately pH 2 (approximately 0.05%) to 90% concentration, specific oxidizing species (e.g., cations like Cu II or Fe III; anions like chromates, nitrates, nitrites) or reducing species (e.g., hydrogen sulfide, stannous salts) can control the redox potential of the sulfuric acid solution. This profoundly influences its corrosive action. Of course, chloride ion contamination also influences specific phenomena, such as pitting or SCC, while fluoride contamination adversely affects the resistance of the reactive metals (i.e., Ta, Ti, and Zr).

Specific Materials

The behavior of specific categories of materials in dilute to intermediate concentrations of sulfuric acid is as follows:

(1) *Light Metals.* Aluminum and its alloys, as well as magnesium alloys, are totally unsuitable for dilute sulfuric acid.

(2) *Iron and Steel.* Cast iron and steel are unsuitable for other than cold concentrated acid. However, nickel cast irons will find some applications in intermediate strength acid, while the high-silicon cast irons (i.e., UNS F47003) are very resistant. However, even the F47003 cast iron will corrode at approximately 20 mpy (0.5 mm/y) in 5 to 55% acid at the atmospheric boiling point. Dilute sulfuric acid can be inhibited against corrosion of steel for chemical cleaning purposes, but the corrosion rates are still too high for long-term exposure.

(3) *Stainless Steels.* Stainless steels must be considered by category. The martensitic and ferritic grades are generally inapplicable in dilute sulfuric, but the type 300 series austenitic grades may sometimes be employed, depending upon the conditions of exposure.

A question that frequently arises is what temperature and concentration limits are relevant in *very* dilute sulfuric (e.g., in pH control of water systems). The data in Table 23.2 for *aerated* acid may be helpful for such decisions.

Table 23.2

Corrosion of Steel in Dilute Sulfuric Acid

% Acid	Temp., °C	Corrosion rate, mpy (mm/y)	
		Type 304	Type 316
0.05	95	0	0
0.25	60	36 (0.9)	0
0.50	90	800 (20.3)	5 (0.1)
1.00	100	300 (7.6)	50 (1.3)

In these very dilute solutions, both dissolved oxygen (a weak passivator) and molybdenum content are significant. No doubt, chloride contamination would adversely influence these rates, which therefore relate to distilled water or demineralized water only. Also, severe corrosion has been observed in the rolled joints of tube-shell heat exchangers, where the dilute acid concentrated locally to some intolerable level due to process-side heat.

In laboratory investigations, 50 to 65% acid at 80 to 90°C ([175 to 200°F] which would otherwise *dissolve* type 304 or 316) has been rendered totally noncorrosive by the addition of as little as 500 ppm Cu II. This would be potentially dangerous in the field, where the inhibitor might not access crevices or joints.

Of the superaustenitic grades, the alloy 20Cb-3[†] (N08020) was specifically developed for dilute and intermediate concentrations of sulfuric acid. The literature contains isocorrosion charts which define the ordinary limitations of concentration and temperature. It must be remembered that these parameters are profoundly affected by specific oxidizing, reducing, or halide contaminants. The newer grades with lower alloy contents (254 SMO[†] [S31254] and AL-6XN[+] [N08367]) will have less resistance than alloy 20Cb-3[†] but more than type 316L.

(4) *Lead*. Lead and its alloys have good resistance to a relatively wide range of sulfuric acid concentrations below the atmospheric boiling point (see isocorrosion charts in the suggested reading materials). However, not only is erosion a problem but the protective sulfate film may be removed by some organic contaminants (e.g., alkyl sulfates, organic acids). On the other hand, an intact sulfate film can render the lead *cathodic* to high alloy valves (e.g., CW-2M [alloy C-4]), the reverse of the initial situation, with attendant galvanic corrosion of the valve unless it is electrically isolated.

(5) *Copper*. The zinc-free copper alloys (e.g., coppers and bronzes) are profoundly influenced by oxidizing and reducing species. They will, in fact, resist dilute sulfuric acid but are corroded by as weak an oxidant as DO. Worse still, the stable corrosion products are the Cu II ions, which are themselves oxidizing species (as opposed to cuprous ions), so the attack is autocatalytic, accelerating with the accumulation of dissolved corrosion products. Copper alloys should only be used in non-oxidizing or reducing conditions.

(6) *Nickel*. Of the nickel alloys, the non-chromium grades like alloy 200 (N02200), alloy 400 (N04400), and B-2 (N10665) resist non-aerated acids free of oxidants. Alloy 400 is similar to copper because of its high copper content (approximately 30%), but is used as auxiliary hardware in acid pickling operations where the hydrogen evolved from steel products keeps the iron and copper salts in a reduced state.

Alloy 600 (N06600) is of little interest in dilute sulfuric, but the molybdenum-bearing grades like alloys 825 (N08825), 625

(N06625), and C-276 (N10276) are useful (manufacturers' literature will have isocorrosion charts).

(7) *Reactive Metals.* Of the reactive metals, titanium is *not* useful in dilute sulfuric acid unless oxidizing contaminants (e.g., Cu II, Fe III, nitrates) are present to maintain passivity.

Zirconium resists up to 70% sulfuric acid at the boiling point, above which concentration rapid "breakaway" corrosion occurs. In some plant operations, it was found that the entire curve derived from laboratory tests in C.P. acid was displaced to the left (i.e., to approximately 60% limiting concentration) by unknown contaminants. Furthermore, *pyrophoric* corrosion products were encountered when a zirconium valve was disassembled. This is a potential ignition hazard where flammable materials (e.g., gases, organic solvents) are simultaneously present.

Tantalum will resist dilute sulfuric acids to the atmospheric boiling point. Only fluoride contamination poses a serious threat, as it does also with zirconium.

(8) *Noble Metals.* Of the noble metals, platinum and gold are quite resistant but see very few industrial applications. The corrosion rate for gold is adversely affected by oxidants. Silver is resistant in very dilute acid, but the rate increases significantly as the concentration or temperature is increased.

(9) *Nonmetallic Materials.* These are routinely used in dilute sulfuric services, with the following caveats.

(a) Glass and other ceramics resist dilute sulfuric acid (in the absence of fluorides). Acid-proof brick construction with a suitable membrane is commonly employed for hot intermediate strength acid vessels.

(b) Rubber and elastomers resist dilute sulfuric acid, within their normal temperature limitations. Organic contaminants might cause problems, as previously discussed.

(c) Of the family of plastics, the fluorinated variety are completely resistant, within their temperature limitations. This is true also of the non-fluorinated types (e.g., polyethylene, polypropylene, PVDC, etc.) *except* that prolonged exposure to, for example, 65% acid at 85°C (185°F) or so will carbonize the plastic due to the oxidizing action of the hot acid. FRP is useful within its temperature limitations, but could also be carbonized in prolonged service at elevated temperatures.

(d) Carbon and graphite are useful materials. Carbon-lined, lead-lined steel pipe has been used for hot intermediate strengths of acid. Carbon brick-lined vessels have been employed. Impervious graphite heat exchangers, although subject to mechanical damage, have many decades of successful applications as calandrias (reboilers) in this type of service.

Handling and Storage

Dilute and intermediate strengths of sulfuric acid are not routinely stored or transported. However, the occasion may arise (e.g., for process day tanks, for regeneration of ion-exchange beds) so some suggestions can be made, subject to the caveats about contamination effects. The recommendations below are intended to resist up to intermediate strengths of sulfuric acid, within their normal temperature limitations, and without undue concern over oxidizing or reducing species in the acid:

Tanks   Brick-lined, glass-lined, plastic-lined, rubber-lined, FRP;

Piping	Plastic-lined, glass-lined, rubber-lined, Pyrex[†], FRP, alloy 625, alloy 20Cb3[†], alloy 625, alloy C-276, zirconium;
Valves	Plastic, rubber, or glass-lined; tantalum-plated, high-silicon cast iron, alloys as for piping above;
Pumps	Same as valve materials; impervious graphite; and
Gaskets	Graphite fiber, felted PTFE, rubber and elastomers.

23.2 Organic Acids

The major organic acids of interest are formic acid (HCOOH), acetic acid (CH_3COOH), and the higher molecular weight acids of the general formula RCOOH (where R indicates an ethyl, propyl, butyl, or other aliphatic group).

23.2.1 Formic Acid

Formic acid is a strong acid, approaching the dilute mineral acids in its activity (i.e., its tendency to release hydrogen ions). It can be particularly aggressive when hot and anaerobic. In the anhydrous state, it is a powerful dehydrating agent. The acid has a tendency to decompose, liberating carbon monoxide and water.

Specific Materials

Following is a brief discussion of specific materials. For more comprehensive details, see the recommended reading materials listed at the end of this chapter:

(1) *Light Alloys.* Although aluminum has been used to ship 95 to 99% formic acid, it is strongly attacked below approximately 30% at temperatures only slightly above ambient.

(2) *Iron and Steel.* Iron and steel are rapidly corroded by formic acid. The 14% silicon grade (UNS F47003) resists water solutions but not anhydrous acid above room temperature.

(3) *Stainless Steels.* Type 304L (UNS S30403) may be used for shipment and storage, but molybdenum-bearing grades are preferred at only slightly elevated temperatures. At elevated temperatures and concentrations, alloy 20Cb-3[†] (N08020) is preferred.

(4) *Lead.* Lead does not resist formic acid (nor do tin or zinc).

(5) *Copper.* Under anaerobic conditions and in the absence of other oxidants, copper and zinc-free copper alloys may be used.

(6) *Nickel.* Alloys 200 (N02200) and 400 (N04400) are not competitive with copper alloys in formic acid. Ni-Cr-Mo grades (e.g., alloy C-276 [N10276]) is used in contaminated acid at elevated temperatures and concentrations.

(7) *Reactive Metals.* Titanium is resistant only in the presence of strongly oxidizing contaminants and can suffer catastrophic corrosion in hot concentrated acid. Zirconium will resist up to 90% acid and tantalum any concentration.

(8) *Precious Metals.* Silver is resistant only to approximately 50% concentration, but platinum resists all strengths of formic acid.

(9) *Nonmetallic Materials.* Glass and ceramic-ware is fully resistant.

Plastics can be attacked because of the solvent nature of formic acid.

Polypropylene may be used to approximately 50°C (120°F), but only fluorinated plastics should be used at elevated temperatures.

Neoprene may be used at ambient temperatures, but fluorinated elastomers are required above room temperature (maximum 50°C [120 °F]).

Carbon and graphite are fully resistant to formic acid.

Storage and Handling

Basically, formic acid is stored and handled in austenitic stainless steel. Heating coils in tanks should be of a high-performance grade, such as alloy 20Cb-3[†] (N08020) and pumps the cast equivalent, CN-7M (N08007).

23.2.2 Acetic Acid

Acetic acid is probably the most commercially important of the organic acids. The specifics of the manufacturing process have a lot to do with the corrosivity in the crude state (although refined or CP [chemically pure] acid is quite predictable). When produced by oxidation processes, peracids or peroxides are formed, which, although unstable, can profoundly affect the behavior of many metals and alloys.

Specific Materials

Behavior of specific categories of materials in acetic acid is as follows:

(1) *Light Metals.* Aluminum is widely used for shipment and storage of refined concentrated acids, free of chlorides or heavy metal ions. (Dilute solutions can be severely corrosive because of increased ionization.) Even in cold concentrated products, a fine haze of aluminum salts may develop, which may be incompatible with product specifications.

(2) *Iron and Steel.* Iron and steel are subject to attack at several hundredths of an inch per year in all concentrations of organic acids and cannot be used for shipment and storage. However, the high silicon iron (UNS F47003) will resist all concentrations at least to the atmospheric boiling point.

(3) *Stainless Steels.* Stainless steels of conventional martensitic and ferritic groups are of no practical importance in this application, but the superferritics (e.g., alloys 29-4C [S44735] and 26-1 [S44626]) can be employed.

Austenitic grades, on the other hand, are very important. Type 304 is used for handling and storage of refined acids up to approximately 70°C (160°F), although "hot-wall" effects can be dangerous and type 316L has become the industry standard. Above that temperature range (or for steam-traced piping), type 316L must be used. Intergranular corrosion is a chronic phenomenon in all hot organic acids. To ensure continued passivity of even austenitic grades, at least trace amounts of oxygen or oxidants must be present. Selective corrosion of welds may be encountered, independently of carbon or sensitization effects, due to microsegregation of molybdenum.

The high-performance alloys (e.g., alloys 825 [N08825] and 20Cb-3[†] [N08020]) handle some very tough borderline conditions. They are sometimes used for welding type 316L vessels or weld-overlaying flange-faces and other areas subject to crevice corrosion.

(4) *Lead.* Lead is not commercially useful in organic acids, although it has sometimes been successfully employed in very dilute acetic acid streams. Oxidizing agents severely aggravate attack.

(5) *Copper.* The resistance of copper and its zinc-free alloys depends entirely upon the presence or absence of oxidizing agents. Copper and high-strength, zinc-free copper alloys are used in all concentrations of organic acids up to the atmospheric boiling point, but only under anaerobic and otherwise non-oxidizing conditions. Even slight attack will severely discolor the refined acid.

(6) *Nickel.* Alloy 200 is rarely employed in this type of service, although alloy 400 (N04400) is occasionally used in lieu of copper. Alloy

600 (N06600) is not usually of any interest in such applications. Alloy B-2 (N10665) is used only for special process conditions. However, alloys 625 (N06625) and C-276 (N10276) are used for the toughest services (e.g., acetic acid vaporizers, seawater cooled exchangers).

(7) *Reactive Metals.* Titanium is the most commonly employed of the reactive metals. It has performed well under oxidizing conditions, but can corrode catastrophically otherwise. Zirconium is an acceptable but more expensive material, while tantalum is resistant but rarely required.

(8) *Noble Metals.* Silver was a traditional material for handling hot organic acids and for heating coils in stainless tanks, but it has been replaced to a great extent by modern nickel-based alloys.

(9) *Nonmetallic Materials.* Plastics are suspect because of the solvent effects of organic acids, but fluorinated grades are fully resistant and polyethylene drums are ideal for ambient temperature storage of C.P. acetic acid. The bisphenol polyesters have been successful in some applications.

Rubber is a traditional material in dilute acetic acid storage. Above 5% concentration, only a few synthetic elastomers (e.g., butyl, EPDM) are resistant, and only butyl rubber will resist glacial acetic up to 80°C (175°F).

Other non-metallics include wood (traditionally used for dilute acetic acid), glass or ceramics, and carbon or graphite. Impervious graphite heat exchangers have been used in the most demanding services for heating or cooling organic acids.

Handling and Storage

The ambient storage of acetic acid presupposes aerobic conditions, because of the high solubility of dissolved oxygen. Following is a listing of materials of construction for a variety of equipment, which selection is thought to constitute good engineering practice with minimum risk:

Tanks	Aluminum or 304L;
Piping	304L;
Valves	CF-3M or CF-8M;
Pumps	CF-3M; and
Gaskets	Spiral-wound PTFE/stainless steel.

23.3.3 Other Organic Acids

Generally speaking, the higher molecular weight acids are less corrosive than acetic acid. The materials recommendations for acetic acid can usually be used for propionic, butyric, and other acids and will be conservative. Special problems are sometimes encountered, as with polymerization of acrylic acid, but do not usually affect metallic materials.

Reference

1. M. G. Fontana, *Corrosion Engineering*, 3rd Ed. (New York, NY: McGraw-Hill, 1986).

Suggested Resource Information

ASM Handbook, Vol. 13, *Corrosion* (Materials Park, OH: ASM International, 1987).

C. P. Dillon, *Materials Selection for Construction for the Chemical Process Industries* (New York, NY: McGraw-Hill, 1992).

W. Z. Friend, *Corrosion of Nickel and Nickel-Base Alloys* (New York, NY: Wiley-Interscience, 1980).

B. J. Moniz, W. I. Pollock, eds., *Process Industries Corrosion–The Theory and Practice* (Houston, TX: NACE International, 1986), pp. 161, 265, 275, and 287.

NACE Technical Report 5A171 "Materials for Receiving, Handling, and Storing Hydrofluoric Acid" (Houston, TX: NACE International).

H. H. Uhlig, R. W. Revie, *Corrosion and Corrosion Control*, 3rd Ed. (New York, NY: Wiley-Interscience, 1985).

Recommended Software Programs

CHEM•COR[†] 2 and CHEM•COR[†] 2 PLUS, Acetic Acid (Houston, TX: NACE International).

CHEM•COR[†] 3 and CHEM•COR[†] 3 PLUS, Formic Acid (Houston, TX: NACE International).

CHEM•COR[†] 4 and CHEM•COR[†] 4 PLUS, Hydrogen Chloride (Houston, TX: NACE International).

CHEM•COR[†] 7 and CHEM•COR[†] 7 PLUS, Phosphoric Acid (Houston, TX: NACE International).

CHEM•COR[†] 9 and CHEM•COR[†] 9 PLUS, Hydrogen Fluoride (Houston, TX: NACE International).

24

Carbon Dioxide

Carbon dioxide is an acidic gas which forms a weak reducing acid (carbonic acid) upon dissolution in water. It is a naturally occurring constituent in air (several hundred parts per million), from which it dissolves in condensed moisture or water, with a buffered pH of approximately 5.7, corresponding to approximately 12 to 15 ppm of CO_2 in the solution. At higher partial pressures of carbon dioxide, as when it is in a natural gas or process steam, a pH as low as 3.5 may be encountered. *Dry* carbon dioxide is noncorrosive.

The weak water solution (carbonic acid) is sometimes thought of as only mildly corrosive. This is because many of the studies have been conducted with natural waters, in which bicarbonates and other ions may buffer the acidity and otherwise influence corrosion.

In distilled water and in process streams containing water, carbonic acid can be severely corrosive to iron and steel. In fact, in carbonic acid, hydrogen evolution begins at approximately pH 5 and *total acidity* is a better indication of corrosivity than is pH.

Low pH solutions will lose carbon dioxide by volatilization at ordinary temperatures and pressures (like a carbonated beverage going flat). Great care must be taken in sampling water solutions for analysis (i.e., containers full to the brim and tightly sealed), to be sure that the samples are truly representative of the original dissolved gases.

24.1 Specific Materials

24.1.1 Light Metals

Magnesium is corroded by mildly acidic carbonic acid solutions, but aluminum and its alloys are resistant (in the absence of heavy metal ions like iron, copper, or lead, and in the absence of chlorides).

24.1.2 Zinc

Zinc, as galvanizing for example, is rapidly corroded by carbonic acid, both because of the acidity and because the corrosion products are quite soluble.

24.1.3 Ferrous Metals

Carbon dioxide alone (e.g., in the absence of dissolved oxygen [DO]) is often reported to be only slightly corrosive (e.g., in natural waters). However, if the carbon dioxide as carbonic acid is constantly replenished (e.g., either by flow or by a high partial pressure of carbon dioxide in the vapor space), corrosion of iron or steel will occur at rates of several tenths of an inch per year. Rates of the order of 200 m/y (5 mm/y) are observed at 25°C (77°F) under 145 psi (1 MPa) CO_2 and at 50°C (122°F) under approximately 30 psi (207 kPa) partial pressure. In the presence of oxygen, there is a synergistic effect, the total corrosion being greater than would be caused by the same amount of DO and carbon dioxide acting separately. Also, cast iron suffers graphitic corrosion in such weak acid environments.

The ASTM A 242[1] steels of improved resistance to atmospheric corrosion (see Chapter 21) are somewhat better than carbon steel in steam condensate contaminated with carbon dioxide.

24.1.4 Stainless Steels

All grades of stainless steels are satisfactorily resistant to carbonic acid. Corrosion, when it does occur, is the result of contaminants, such as chlorides, or the conjoint or independent action of other species such as hydrogen sulfide, ammonia, or low-boiling alkyl amines.

24.1.5 Lead

Lead and its alloys are usually only slightly attacked by carbonic acid solutions, except under velocity conditions. However, corrosion is severe in the presence of oxygen, or particularly, peroxides or other strong oxidants.

24.1.6 Copper Alloys

Except for the problem of dezincification of yellow brass, corrosion of copper by carbonic acid is controlled solely by DO or other oxidants. In practice, copper alloys are rarely employed, because of the ever-present danger of oxygen ingress and the economic superiority and reliability of other materials, such as stainless steels.

24.1.7 Nickel Alloys

With the exception of alloy 400 (UNS N04400), which is attacked at certain critical ratios of oxygen to carbon dioxide in water, all nickel alloys resist attack by carbonic acid. However, they are not usually an economical choice unless there are other corrosive species, such as chloride, present simultaneously.

24.1.8 Reactive and Noble Metals

With the possible exception of titanium (under low-pH), chloride-contaminated and anaerobic conditions, such materials (while resistant) would be uneconomical for this service.

24.1.9 Nonmetallic Materials

All plastics and elastomers of industrial interest are resistant to carbonic acid corrosion within their normal temperature/pressure limitations.

Cement and concrete are subject to severe attack by carbonic acid, due to the solubilization of calcium compounds. However, glass, ceramics, carbon, and graphite are inert.

Reference

1.   ASTM Standard A 242, "Specification for High-Strength Low-Alloy Structural Steel" (Philadelphia, PA: ASTM).

25

Hydrogen Sulfide

Hydrogen sulfide (H_2S) is a colorless, flammable gas which has a "rotten egg" odor at low concentrations. It is *lethal* (more so even than hydrogen cyanide). It is particularly dangerous because it paralyzes the olfactory nerves at a relatively low concentration. Persons are often unaware of their exposure to dangerous concentrations of H_2S, and fatalities have occurred from exposure to sour gas streams, "poisonous dawn fogs" in swamplands, or from marine eruptions which release hydrogen sulfide from bacterial action.

As a corrosive species, hydrogen sulfide is of concern in many oil, gas, and petrochemical operations. Hydrogen sulfide is a decomposition product of many sulfur-bearing compounds and certain natural organic nutrients. While it is often an undesirable constituent to be removed from a product (e.g., natural gas), it may also be a desired reactant (as a feedstock for the production of sulfur; as an extractant for deuterium oxide from water in the production of heavy water, for example).

25.1 Corrosivity

25.1.1 General Corrosion

A weak acid in water solution (e.g., 0.3%; approximately three times as soluble as CO_2), and having a pH of approximately 5.5, it is only moderately corrosive to steel when acting as the sole corrosive species. However, should the protective iron sulfide film be solubilized or swept away by velocity effects, fresh metal is continuously exposed to attack and high corrosion rates are experienced.

In combination with dissolved oxygen (DO), hydrogen sulfide can be highly corrosive, often resulting in pitting corrosion or formation of complex reaction products (e.g., polythionic acids) which are corrosive. In other circumstances, hydrogen sulfide may react with DO to form polysulfides, which may result in protective iron polysulfide films.

Corrosion is increased in the presence of carbon dioxide, although this may be due simply to the higher total acidity experienced from the combined acid gases.

25.1.2 Localized Corrosion

In oil and gas services, a variable degree of protection from oil-saturated corrosion products will lead to the formation of corrosion "lakes," shallow sites of localized corrosion surrounded by film-protected areas. If small imperfections exist in an otherwise protective iron sulfide film, a severe pitting type of attack may be experienced.

25.1.3 Hydrogen Effects

A most unfortunate characteristic of hydrogen sulfide is its ability to inhibit the dimerization of nascent atomic hydrogen (formed at the cathodes by corrosion) to molecular hydrogen (see also Chapter 30). The atomic hydrogen readily penetrates the metal lattice, as previously described, and this effect may be aggravated by iron sulfides.

Steel may suffer slow-strain rate embrittlement (SSRE), rendering it susceptible to brittle fracture or may develop *hydrogen blisters*. Lastly, sulfide stress cracking (SSC), stress-oriented hydrogen-induced cracking (SOHIC), or step-wise cracking (SWC) in piping, may develop.

25.1.4 Environmental Cracking

Anodic Cracking

SCC of austenitic stainless steels (e.g., types 304L and 316L) can occur in the simultaneous presence of sulfide and chloride ions, and at lower chloride concentrations and much lower temperatures than when sulfide is absent. This effect is most prevalent in heavy water manufacturing processes. The sulfides apparently compete with oxygen at adsorption sites, locally weakening the passive film. This sulfide-chloride combination results in transcrystalline cracking, even at room temperature, in the presence of only a few hundred ppm of chloride ion. Cracking is often initiated at hard weld-heat-affected zones (HAZs), but may also occur in relatively soft base metal.

Cathodic Cracking

Hydrogen sulfide is specific for SSC of hardened steel and other alloys, as previously described.

25.2 Specific Materials

The effects of wet hydrogen sulfide on the various categories of materials (*dry* gas is noncorrosive up to approximately 250°C [480°F]) is as follows.

25.2.1 Aluminum

Aluminum and its alloys will withstand dry or wet hydrogen sulfide, even at elevated temperatures. However, oxidation of hydrogen sulfide will form H_2O and SO_2 (sulfurous acid), which *is* potentially corrosive.

25.2.2 Zinc

Because of its poor resistance to acidic conditions, zinc (e.g., as galvanized steel) has little application in hydrogen sulfide-bearing environments. (*Note*: Zinc-pigmented prime coats contaminated by hydrogen sulfide-polluted atmospheres do not accept properly the application of a top-coat paint.)

25.2.3 Iron and Steel

In the absence of oxygen and below pH 5.5, corrosion of iron and steel greatly increases with increasing acidity. At pH >5.5, hydrogen sulfide begins to dissociate to form bisulfide ions. Corrosion by bisulfides increases with pH in the 5.5 to 7.5 range until a protective iron sulfide film is formed above pH 7.5. High corrosion rates are usually accompanied by hydrogen embrittlement (HE), blistering, and various forms of cracking. When undisturbed protective sulfide films are present on iron and steel, corrosion rates are typically less than 0.5 mm/y (20 mpy). There is, however, a tendency towards pitting at weak points in the film, and the presence of DO can increase the pitting rate approximately thirty-fold. Slow strain rate embrittlemnt (SSRE) is a potential problem (e.g., in the vapor space of tanks holding sour petroleum products).

Hydrogen blistering is reduced by using clean steels (i.e., steels of fine grain size, having a maximum 0.006% sulfur content with inclusion shape control or 0.002% without). Hardened and highly stressed steels, are subject to SSC.

25.2.4 Stainless Steels

The martensitic grades (e.g., type 410 [UNS S41000]) are resistant to aqueous H_2S but subject to SSC precisely because they tend to be used in the hardened condition. Ferritic grades (e.g., type 430 [S43000]) are more resistant and the superferritics (e.g., alloy 29-4C [S44735], and 26-1 [S44627]) are considered to be fully resistant. (*Note*: The maximum allowable hardness for these and other alloys is defined in the NACE Standard MR0175.[1])

The austenitic grades (e.g., types 304 and 316) are immune to both corrosion and SSC (unless *severely* cold-worked to a hardness in excess of HRC 22 [Rockwell C 22]) but are subject to the sulfide/chloride transcrystalline stress cracking described above. High-performance stainless steels and nickel-rich alloys described in Chapter 13 are immune to this effect.

When type 300 series stainless steels are *sensitized*, they are subject to the intergranular SCC known as *polythionic cracking*. Polythionic acids (the tetrathionic acid and its salts are thought to be the worst) are formed by the interaction of wet hydrogen sulfide with sulfur dioxide, and water solution formed in this manner is used in laboratory tests. In plant, the polythionic acids are apparently formed from moisture, atmospheric oxygen, and certain iron/chromium sulfide scales. To combat this type of attack during down-time, refinery operations usually specify a stabilized grade of stainless steel (e.g., alloys 347 [S34700] and 321 [S32100]), which is substantially immune to sensitization and weld decay. NACE Standard RP0170 recommends specific shutdown procedures to prevent polythionic cracking of sensitized stainless steels.[2]

Precipitation-hardening grades (e.g., 17-4 PH [S17400] and 17-7 PH [S17700]) require specific heat treatment control of hardness levels for sour service, per NACE MR0175.

25.2.5 Copper Alloys

Copper and the various bronzes have poor resistance to wet hydrogen sulfide, because of the voluminous non-protective black sulfide corrosion products. However, *yellow* brasses (e.g., admiralty B [C44300] and aluminum brass [C68700]) can be quite resistant, simply because their zinc content produces a dense, adherent black tarnish. Normally, use of yellow brasses is not recommended in sour service, although they have been successfully used in specific applications such as refinery coolers and condensers. Cupronickels suffer localized pitting in salt waters contaminated with hydrogen sulfide (e.g., as from SRB attack).

25.2.6 Nickel Alloys

Although chromium-free nickel alloys are more resistant than copper alloys, only the chromium-bearing alloys are highly resistant (e.g., alloys 600 [N06600], 625 [N06625], and C-276 [N10276]). Even these can suffer SSC *if* highly cold-worked and the cathodic member of a galvanic couple, where nascent hydrogen is generated.

On all nickel alloys, sulfur contamination may lead to LMC if heated above approximately 600°C (1,100°F) or welded. This is due to the low-melting nickel sulfide eutectic previously described.

25.2.7 Lead

Lead and its alloys are moderately resistant to hydrogen sulfide in the absence of velocity or abrasive effects.

25.2.8 Reactive Metals

Although titanium, zirconium, and tantalum, are extremely sensitive to hydrogen pick-up, they are inherently resistant to wet hydrogen sulfide *itself* at temperatures below roughly 150°C (300°F). They could suffer hydriding when constituting cathodes in galvanic couples. However, titanium reportedly is an excellent material for instrumentation in certain specific sour services, and has been used extensively in refinery coolers and condensers.

As a possible contaminant in more corrosive acids, hydrogen sulfide would act as a reducing agent, with possible adverse effects on the corrosion resistance of titanium.

25.2.9 Noble Metals

Gold is resistant, and platinum very much so, over a wide range of temperatures and concentrations. Silver, however, is moderately attacked. The commonly observed "tarnishing" of silver is due specifically to traces of hydrogen sulfide in the atmosphere.

25.2.10 Nonmetals

Of the nonmetals, only rubber and elastomer seem to pose any problem. They are subject to vulcanization and chemical deterioration in the presence of hydrogen sulfide at elevated temperatures and pressures (e.g., in down-hole oilfield applications).

25.3 Pyrophoric Products

Certain metals can form pyrophoric corrosion products. The most common is pyrophoric iron sulfide, probably ferrous disulfide. It is not definitely known whether this forms directly on steel or indirectly from other corrosion products. It may require some oxygen present (to oxidize ferrous sulfide), or perhaps pre-existing rust products (e.g., FeO[OH], ordinary ferric oxide). The specific problem is that these pyrophores get red-hot on exposure to air or oxygen (e.g., when equipment is opened for maintenance) and are a dangerous ignition source if flammable gases or liquids are about. Tires on vehicles have also caught on fire when parked on deposits scraped out of sour service pipelines.

References

1. NACE Standard MR0175, "Sulfide Stress Cracking-Resistant Metallic Materials for Oilfield Equipment" (Houston, TX: NACE International).
2. NACE Standard RP0170, "Protection of Austenitic Stainless Steel from Polythionic Acid Stress Corrosion Cracking During Shutdown of Refinery Equipment" (Houston, TX: NACE International).

Suggested Resource Information

R. N. Tuttle, R. D. Kane, *H₂S Corrosion in Oil & Gas Production–A Compilation of Classic Papers* (Houston, TX: NACE International, 1981).

26

Corrosion by Chlorine

The corrosion behavior of chlorine is profoundly affected by the presence or absence of water. Dry chlorine as either liquid or vapor is a powerful oxidant and chlorinating agent. Water solutions are aggressive oxidizing acids, containing both chloride and hypochlorite anions. While dry chlorine is not particularly corrosive at ambient temperatures (it is stored and handled in ordinary carbon steel), it becomes increasingly active with increasing temperature. Some metals will burn readily in dry chlorine (e.g., titanium).

26.1 Specific Materials

26.1.1 Light Metals

Theoretically, aluminum and magnesium will resist dry chlorine. In practice, they are not considered for such service because they will react rapidly if even traces of moisture should become available in the system.

26.1.2 Iron and Steel

Iron and steel will resist dry chlorine to temperatures of approximately 175°C (345°F) and 200°C (390°F), respectively. Liquid chlorine is routinely shipped, stored, and vaporized in steel equipment. Any ingress of water will result in rapid and severe corrosion. Traces of ferric chloride are inevitably present, which may clog fine orifices and interfere with valve seating.

26.1.3 Stainless Steels

The stainless steels will resist dry chlorine up to approximately 300°C (570°F). However, they are severely attacked if there is ingress of water. Also, being subject to chloride pitting and SCC with even parts per million of water, they are not usually employed in chlorine services. SCC failures have occurred in stainless fasteners in swimming pool ceilings, due to wet chlorine vapors.

26.1.4 Lead, Zinc and Tin

Lead will withstand both wet and dry chlorine up to approximately 110°C (230°F). It has been used successfully as a gasket material in such applications. Since it depends for its resistance on a film of insoluble corrosion products, it is not employed where significant flow or turbulence would be encountered.

Zinc and tin are not normally employed in chlorine service.

26.1.5 Copper Alloys

Copper and its alloys will withstand dry chlorine up to approximately 200°C (390°F). However, there is little reason to select them for such service and they would be severely attacked if any ingress of moisture should occur.

26.1.6 Nickel Alloys

Although all but the nickel-chromium-molybdenum alloys (e.g, alloy C-276 [UNS N10276]) are attacked by wet chlorine, nickel and its alloys have excellent resistance to *dry* chlorine, even at elevated temperatures. The temperature limitations for some common nickel alloys are listed in Table 26.1.

Table 26.1

Nickel Alloys in Chlorine Service

Alloy	UNS	Temperature Limits	
		°C	°F
200	N02200	400	750
201	N02201	535	995
400	N04400	425	798
600	N06600	535	995
B	N10001	535	995
C-276	N10276	510	950

26.1.7 Reactive Metals

Titanium will not resist dry chlorine. A minimum of 1,500 ppm water must be present or ignition will occur. Although titanium has outstanding resistance to wet chlorine, there is an inherent danger in crevices, in which the moisture may be consumed or the critical balance of hydrochloric to hypochlorous acid be disturbed. Serious crevice corrosion may then ensue.

Zirconium offers no advantage in chlorine service. Tantalum is useful in both wet and dry chlorine up to approximately 150°C (300°F).

26.1.8 Plastics

Only the fluorinated plastics can be exposed to both wet and dry chlorine. In chemical processes, even PTFE may be unsuccessful because of chlorination of other organic materials absorbed within the interstices of the PTFE. This will cause mechanical damage (e.g., of valve diaphragms), even though the PTFE is itself unattacked.

PVC is widely used to handle wet chlorine (and dry chlorine well diluted with other gases). In concentrated chlorine, however, the PVC will continue to chlorinate and degrade.

Polyesters must not be exposed to dry chlorine but will withstand wet chlorine to approximately 90°C (195°F). Vinyl polyesters are preferred.

26.1.9 Rubber and Elastomers

Hard natural rubber has long been used in wet chlorine service. In general, rubber and elastomers must not be exposed to dry chlorine; even the fluorinated elastomers suffer attack of certain constituents.

26.1.10 Other NonMetals

Glass and other ceramic materials are unaffected by either wet or dry chlorine.

Carbon and graphite will withstand dry chlorine to approximately 1,650°C (3,000°F) in the absence of oxygen. Otherwise, they can be used only to approximately 330°C (625°F) and 425°C (795°F), respectively. Impregnated impervious graphite will tolerate dry chlorine to approximately 120°C (250°F), but wet chlorine attacks the organic binders.

26.2 Storage and Handling

There should be no occasion for storing wet chlorine, but glass-lined, titanium, or alloy C-276 (N10276) equipment would be resistant. Dry chlorine is stored, with minimum risk, as indicated below:

Tanks	Steel;
Piping	Steel;
Valves	Steel with alloy 400 trim and alloy C-276 stems;
Pumps	Ductile iron, Ni-resist, alloy 400; and
Gaskets	Spiral-wound alloy 400 and PTFE.

Suggested Resource Information

Handbook for Piping Systems–Cl$_2$, ClO$_2$, etc. (Midland, MI: Dow Chemical Company).

D. W. F. Hardie, *Electrolytic Manufacture of Chemical from Salt* (Washington, DC: The Chlorine Institute, 1975).

D. R. Hise, "Corrosion from the Halogens," *Process Industries Corrosion* (Houston, TX: NACE International, 1975), p. 240.

B. J. Moniz, W. I. Pollock, eds., *Process Industries Corrosion–The Theory and Practice* (Houston, TX: NACE International, 1986).

Recommended Software Program

CHEM•COR[+] 5 and CHEM•COR[+] 5 PLUS, Chlorine (Houston, TX: NACE International).

27

Corrosion in Alkaline Environments

Pre-eminent among the alkaline environments are sodium hydroxide and ammonia (the latter is covered, together with its organic derivatives, the amines, in the next chapter). Sodium hydroxide is commonly called caustic soda. The term *caustic* refers to materials capable of eating away (or burning, especially as to flesh); it is loosely used for the group of strong alkalis like sodium hydroxide, potassium hydroxide (caustic potash), and calcium hydroxide (caustic lime). Sodium carbonate is a relatively strong alkali, but its major problem is a tendency to hydrolyze to sodium hydroxide at elevated temperatures.

27.1 Sodium Hydroxide

27.1.1 Properties

Sodium hydroxide (NaOH) is a white hygroscopic solid with a melting point of 318°C (604°F) and a solubility in water at room temperature which yields a 78% concentration.

With amphoteric metals (i.e., those attacked by both acids and alkalis, such as aluminum or zinc), the alkali reacts directly with the metal, liberating hydrogen, just as in acid attack. With other metals and alloys, reaction is characterized by the formation of relatively insoluble hydroxides on the surface, limiting the attack. For this reason, caustic or alkaline materials are not as universally corrosive as acids.

Corrosion, when it occurs, is due to special effects, such as the formation of complex compounds, double salts, basic salts, or double hydroxides. An example is the reaction of iron with caustic to form sodium ferroate. As the caustic solutions become hotter and more concentrated, there is a greater tendency for them to react directly with metal. In effect, iron and steel exhibit amphoteric characteristics in strong, hot sodium hydroxide.

27.1.2 Specific Materials

The following sections deal with resistance of the various groups of material to alkaline environments.

Light Metals

Aluminum and magnesium are non-resistant to caustic.

Iron and Steel

The corrosion resistance of iron and steel is usually acceptable up to approximately 70% caustic at 80°C (175°F), except for the problem of SCC (i.e., "caustic embrittlement"), if iron contamination is not objectionable. However, the corrosion rate increases rapidly with temperature, reaching more than 12.5 mm/y (500 mpy) at 100°C (212°F). This can cause high rates with steam-traced carbon steel piping, for example, when static rather than flowing conditions prevail.

Caustic embrittlement is a misnomer; the metal matrix retains its inherent ductility, although a brittle fracture typical of environmental cracking ensues. The parameters of temperature and concentration above which SCC is expected to occur are illustrated in Figure 27.1.

The time to failure can be greatly extended by thermal stress-relief, which is mandatory for either *welded* or *cold-formed* steel equipment (e.g., flared pipe), within the SCC limits. *Applied* stresses are also to be minimized. It is generally believed that stresses must be at or above the yield point for caustic cracking to occur. Because of the relatively high freezing point of concentrated sodium hydroxide (which necessitates heating lines and vessels), SCC is a problem even under ordinary storage conditions.

Additions of *nickel* to iron and steel greatly increase corrosion resistance. The high nickel cast irons (e.g., the several grades of Ni-Resist[†]) are popular for caustic services.

Stainless Steels

The conventional straight chromium grade martensitic and ferritic stainless steels are unreliable in caustic above ambient

[†] Trade name

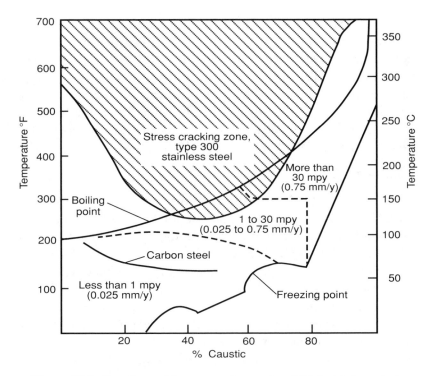

Figure 27.1 Caustic cracking curves for steel and 18-8 stainless steel.

temperatures. They may become active and exhibit less resistance than ordinary steel.

The superferritics (e.g., alloy 26-1 [UNS S44626]) have distinctly different properties and have been used successfully in caustic evaporators.

Note: This application may be profoundly influenced by the concentration of residual oxidizing species, such as chlorates, in the caustic. At temperatures greater than approximately 300°C (570°F), SCC is a real problem.

The type 300 series is more resistant than ordinary 12% and 17% chromium grades, but can also lose passivity and suffer severe corrosion. The probable practical limits for types 304L and 316L are 50% caustic at 70°C (160°F) and 40% at 80°C (175°F).

At and above 300°C (570°F), the austenitic stainless steels including the superstainless grades, are easily stress-cracked by caustic, or even by sodium carbonate, because of hydrolysis effects. Sodium hydroxide of 10% concentration has caused SCC in autoclave tests. Plant failures are encountered in environments ranging from caustic-contaminated steam to molten sodium contaminated concentrations of water in parts per million. The temperature at which SCC of type 300 series stainless steel by caustic becomes a problem is not known with certainty, and probably varies with oxygen/oxidant content. A curve based on laboratory data in chemically pure sodium hydroxide is included in Figure 27.1, page 279 (the curve for steel is based on plant experience).

Copper Alloys

There is relatively little information about copper alloys in the literature, because copper contamination is objectionable in the rayon industry (color) and the soap industry (rancidity), both of which are major users of caustic. Nevertheless, in the absence of strong oxidants (e.g., chlorates, hypochlorites), copper alloys are very resistant. The cupronickels have been used in many applications in which copper contamination was not a problem. Low-zinc bronze valves and pumps are very economical for handling caustic (oxidant-free), while copper pipe has sometimes been used when it was impractical to stress-relieve steel pipe in the field.

Nickel Alloys

Commercially pure nickel (i.e., alloy 200 [UNS N02200]) will handle any and all caustic solutions and even molten 100% sodium hydroxide. Eventually, it may suffer SCC at temperatures of the order of 325°C (620°F) over a prolonged period of several years. It is widely used, in the form of nickel-clad steel vessels, to handle strong hot caustic. Above approximately 315°C (600°F), the low-carbon variety (i.e., alloy 201 [N02201]) is recommended.

Alloy 600 (N06600) is often preferred for piping, because it is more easily welded in the field. It is also preferred for heating coils, because of its greater strength at steam temperatures, although this may introduce trace amounts of hexavalent chromium ions (which may be detrimental to produce quality). SCC will eventually occur at temperatures greater than 300°C (757°F), but it takes months or years to develop (compared to within hours for type 300 series stainless steels).

Alloy 400 (N04400) is somewhat less resistant than alloy 600, and more susceptible to SCC by a factor of three or more.

Furthermore, its iron and copper content make it objectionable for the more meticulous applications.

The higher nickel alloys (e.g., alloys B-2 [UNS N10665] and C-276 [N10276]) are rarely needed or considered for this type of service, but should be highly resistant.

Reactive Metals

Titanium has good resistance up to approximately 40% caustic at 80°C (175°F). Resistance is improved by anodizing the metal to reinforce the surface oxide film, and also by the presence of oxidizing species in the caustic. Galvanic couples may cause hydriding, and chloride contamination may lead to penetration of the protective oxide film at elevated temperature.

Zirconium resists attack in almost all alkalis, either fused or in solution. The temperature limit in 70 to 73% caustic is approximately 130°C (265°F).

Tantalum is severely attacked by caustic.

Noble Metals

Silver has excellent resistance to even very hot, concentrated caustic. It was the traditional material of construction prior to the development of the nickel alloys. Gold and platinum are fully resistant, but have no common industrial application.

Non-Metals

Carbon and graphite are inherently resistant up to the atmospheric boiling point in all concentrations of caustic. However, commercial impervious graphite will suffer attack on the organic impregnant, and the use of cemented joints further limits its applicability. Phenolic resins are non-resistant, and even with epoxy resins applications are limited to 10% caustic (maximum) and approximately 125°C (255°F).

Glass is readily attacked by hot caustic, the temperature being more limiting than concentration. Glass-lined equipment may be used for low temperatures and concentrations within the parameters shown in Figure 27.2, which has been used to set operating limits for such equipment. Assuming an 80-mil (2-mm) thickness, for example, and allowing a 100% margin of error, an eight-year

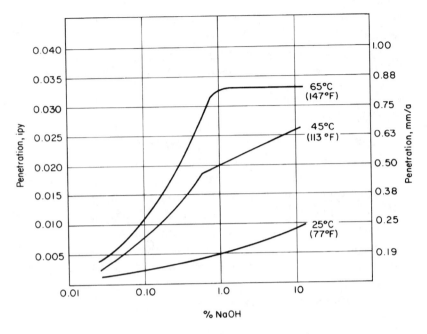

Figure 27.2 Caustic vs borosilicate glass.

life is limited by 0.1% caustic at 65°C (150°F) and a thirteen-year life by 0.5% caustic at approximately 50°C (120°F).

Unreinforced plastics (e.g., PVC, PE, PP) are suitable for caustic solutions, within their normal limitations of temperature and pressure.

FRP piping based on epoxy resins has been used for up to 50% caustic, within its temperature limitations. An incremental advantage is gained by incorporating a Dynel[†] veil below the surface of the gel coat, to protect the fibers of the glass reinforcement from direct contact with alkali.

27.1.3 Handling and Storage

Following is a listing of materials of construction for various items of equipment, which listing is thought to constitute good engineering practice with minimum risk:

Tanks	Carbon steel, stress-relieved;
Piping	Carbon steel, stress-relieved, cast iron; FRP (epoxy), plastic-lined steel, alloy 600;

Valves   Ductile iron; low-zinc bronze; steel (stainless trim);

Pumps   Ductile iron, bronze, nickel cast iron, CF-3M; and

Gaskets   Elastomeric; Graphoil[†].

27.2 Potassium Hydroxide

In general, aqueous solutions of caustic potash are considered to have the same corrosion characteristics as caustic soda solutions of the same concentration.

27.3 Calcium Hydroxide

Despite its strong effect on flesh and other organic materials, lime solutions are not particularly corrosive to metals, other than those with amphoteric properties. This is due to a low and inverse solubility (approximately 1,500 ppm in water at room temperature, diminishing with increasing temperature). Lead has been attacked by lime-water freshly leached from cement.

27.4 Alkaline Salts

Alkaline salts are products of a strong base (e.g., sodium hydroxide) and a weak acid (e.g., carbonic acid, acetic acid, phosphoric acid). They act like a weak alkali. The alkaline salts are those of sodium, potassium, lithium, calcium, magnesium, and barium with weak acids. They are discussed in more detail in Chapter 29.

Suggested Resource Information

B. J. Moniz, W. I. Pollock, eds., *Process Industries Corrosion–The Theory and Practice* (Houston, TX: NACE International, 1986), p. 297.

Recommended Software Program

CHEM•COR[†] 6 and CHEM•COR[†] 6 PLUS, Sodium Hydroxide (Houston, TX: NACE International).

28

Ammonia and Its Compounds

Ammonia is a pungent gas which dissolves in water to form the alkaline ammonium hydroxide. Ammonium salts are encountered particularly in the fertilizer industry. Organic derivatives are called *amines*, the bulk of which are encountered as alkyl amines, alkylene amines such as ethylene diamine, or alkanolamines such as monoethanoamine (MEA).

28.1 Ammonia and Ammonium Hydroxide

Anhydrous ammonia, as explained in Chapter 8, is specific for SCC of carbon steel unless inhibited with approximately 2,000 ppm water. It is routinely handled in bulk storage in steel, however. The presence of moisture or the formation of an aqueous solution changes the corrosion characteristics considerably.

28.1.1 Specific Materials

The behavior of specific groups of materials to ammonia and ammonium hydroxide are as follows.

Light Metals

Aluminum and its alloys are mildly corroded (less than approximately 0.1 mm/y) in ammonium hydroxide up to approximately 50°C (120°F).

Iron and Steel

Cast iron and steel are suitable for wet ammonia and ammonium hydroxide if iron contamination is not a problem. Rusting will occur, particularly in the vapor space of tanks, but the attack diminishes with time as the vessel becomes "seasoned." In ammonia with inadequate water content, SCC of older vessels, if not thermally stress-relieved, can occur at ambient temperature, because of uneven distribution between the liquid and vapor phases of the water as a corrosion inhibitor.

Stainless Steels

All grades of stainless steel resist ammonia and ammonium hydroxide at ambient temperatures and moderately elevated temperatures (certainly up to 100°C [212 °F]).

Copper and Its Alloys

Copper alloys are usually not employed in ammonia services, both because of the SCC problem and because of corrosion by formation of the copper-ammonia complex, as indicated by the formation of royal blue-colored corrosion product. The latter, however, requires the presence of oxygen or other oxidizing agents for its formation.

Nickel and Its Alloys

Nickel and its non-chromium alloys are more resistant than copper, and immune to SCC, but also form colored corrosion products in the presence of oxidants. They are rarely employed in ammoniacal services. The chromium-bearing alloys (e.g., alloy 600 [UNS N06600]) offer no advantage over type 300 series stainless steels in such applications.

Noble and Reactive Metals

Within these groups, silver can form very dangerous *explosive* complexes (azides) with ammonia and its derivatives. Gold and platinum, or the reactive metals, find no application in ammoniacal services. Tantalum (e.g., as patches in glass-lined equipment) would be suspect because of its low alkali resistance.

Nonmetals

Siliceous ceramics are attacked by ammonium hydroxide. Carbon and graphite are inherently resistant, but phenolic binders are attacked. Plastics and elastomers resist ammonia solutions within their normal temperature/pressure limitations.

28.1.2 Special Problems

A little-known problem arises from the reaction of ammonia with carbon dioxide. In the presence of large amounts of water, the noncorrosive ammonium carbonate or bicarbonate is formed. However, at slightly elevated temperatures (e.g., 60°C [140°F]) and atmospheric pressure or above, trace amounts of a corrosive species are formed (even in as dilute a medium as 20% ammonium carbonate, or in the vapors over ammonium hydroxide). This species is probably ammonium carbamate or a derivative, and can attack steel (and even stainless steel at higher temperatures and pressures) at rates of the order of 3 mm/y (120 mpy) or more. This is the probable cause of a number of corrosion failures in vessels stripping ammonia from a process environment, when the potential problems due to carbon dioxide ingress were not recognized in advance.

28.2 Amines

Certain characteristic problems arise from the handling of amines, either as the product itself or in acid-gas scrubbing systems used to remove carbon dioxide and/or hydrogen sulfide from other gases. The problems arise from the dual nature of amines, which are both strong organic solvents and (in water solution) alkalis.

28.2.1 Specific Materials

The corrosion resistance of specific groups of materials to amines is as follows.

Light Metals

Aluminum is commonly used for the shipment and storage of refined amines. However, it must not be used as a *heating coil* in hydroxylated amines (e.g., monoethanolamine) because of severe exothermic corrosion after an induction period of some indeterminate time.

Iron and Steel

Amines of all kinds readily form iron complexes with steel and may not be stored in steel when iron contamination is objectionable. Steel tanks are employed for MEA and other such products to be used in acid-gas scrubbing systems. Amines can be severely corrosive to steel under process conditions at elevated temperatures, particularly above 80°C (175°F). Amines are not specific for SCC. When encountered, cracking is usually due to contaminants, notably hydrogen sulfide, which causes SSC of critically hard-steel components (see Chapter 25), although cyanides in acid gas streams have also caused cracking.

Stainless Steels

Amines are successfully stored in type 304 tankage, but the molybdenum-bearing grades (e.g., type 316) are required above 100°C (212°F) in their manufacturing process. Intergranular corrosion, requiring type 316L, is not usually encountered unless there are specific contaminants present. Valves and pumps should be CF-8M or equivalent, rather than CA-6NM (J91540) or other lower alloys.

Lead

Lead is subject to corrosion by aqueous amines. Existing lead-lined tanks have been used for storage of MEA at ambient temperatures, when lead contamination was not objectionable.

Copper and Its Alloys

As with ammonia, copper alloys can resist amines under anaerobic conditions. However, the ever-present danger of color formation, corrosion, and SCC (e.g., due to ingress of air and moisture) renders them generally unacceptable except for certain well-defined and controlled conditions.

Nickel and Its Alloys

With a few exceptions, and because of cost considerations, nickel alloys are only rarely selected for amine-type services. Alloy C-276 (UNS N10276) centrifuges have been used, but only to combat specific process problems such as in urea plants or in amine manufacture.

Nonmetallics

Carbon and graphite are inherently resistant, but the solvent action will attack phenolic impregnants. Epoxy-cemented impervious graphite heat exchangers are successfully used in acid-gas scrubbing systems such as for MEA solution interchangers.

Suggested Resource Information

C. P. Dillon, *Materials Selection for the Chemical Process Industries* (New York, NY: McGraw-Hill, 1992).

B. J. Moniz, W. I. Pollock, eds., *Process Industries Corrosion–The Theory and Practice* (Houston, TX: NACE International, 1986), p. 311.

29

Corrosion by Salts

Salts are the products of reaction between an acid and a base (e.g., sodium chloride from caustic and hydrochloric acid). In this case, the salt is a neutral salt, having neither acidic nor base characteristics because it is the product of a strong acid and a strong base. Water solutions of neutral salts simply increase the electrical conductivity of the solution, although the effect of specific ions such as chlorides causing pitting or SCC must also be considered. In some cases, an oxidizing anion will also have specific effects, as with sodium nitrite, for example.

Acid salts are formed by the interaction of a strong acid and a weak base, and act very much like a weak solution of the parent acid. For example, ammonium sulfate solutions act like dilute sulfuric acid except for reactions specific to the ammonium ion, as with copper. Also, oxidizing cations will affect the redox potential and the corrosion characteristics of the solution, as with cupric sulfate or ferric chloride, for example.

Basic salts are the product of a strong base (e.g., sodium hydroxide) and a weak acid (e.g., acetic acid). They act like a weak alkali. Any oxidizing capacity will derive from the acidic anion (e.g., as in sodium chromate, sodium nitrite, or sodium hyphochlorite), as there are no strong oxidizing alkalis.

Oxidizing salts, whether acidic or basic, pose special problems and are discussed separately below.

29.1 Neutral Salts

These are usually sodium or potassium salts formed from the corresponding alkali. The sodium or potassium ions per se usually cause problems only in high temperature corrosion (Chapter 31). The anions do have specific effects, as with chlorides causing pitting or SCC, and sulfates being reduced by bacterial action to sulfides and dissolved sulfur.

Increasing concentrations of salt lower the solubility of dissolved oxygen. Note that a solution with 6,000 ppm of chloride ion in water has maximum corrosivity to steel, because of lower dissolved oxygen (DO) at higher concentrations.

The increased conductivity associated with any salt solution may aggravate galvanic corrosion by allowing greater cathode-to-anode ratios to come into play.

Neutral sulfates are aggressively corrosive to portland cement mixtures, as previously discussed.

29.2 Acid Salts

Although these act like weak solutions of the parent acid, corrosivity diminishes with polybasic acids containing more than one hydrogen, as the hydrogen ions are successively replaced with less acidic cations. For example, monosodium phosphate is more corrosive than disodium phosphate. Trisodium phosphate is a mildly alkaline salt, as discussed further below.

Nonoxidizing acid chlorides (e.g., magnesium chloride) are not only corrosive to steel and other anodic materials by virtue of their acidic nature, but are highly specific for the SCC of type 300 series stainless steels, particularly.

Ammonium salts, in the presence of oxygen or oxidants, aggravate the corrosion of copper and nickel alloys (other than the chromium-bearing variety) by complexing and solubilizing the surface films which would otherwise limit corrosion.

29.3 Alkaline Salts

The alkaline salts are those resulting from the reaction of a strong base (e.g., sodium, potassium, lithium, etc.) with a weak acid. They act like a weak alkali. In most cases, the sodium salts are our primary concern.

Sodium carbonates and bicarbonates, particularly, may be thermally decomposed to yield free caustic, with its attendant problems. Otherwise, the sodium salts are a problem primarily with materials not resistant to alkalis. Trisodium phosphate, for example, is a mildly alkaline compound whose water solution is noncorrosive to copper and steel but has a mild etching effect on aluminum.

If the anion has oxidizing capacity, the corrosion characteristics can be drastically different.

29.4 Oxidizing Salts

Neutral or alkaline oxidizing salts, if they contain no halides as constituents or contaminants, are not very corrosive. Sodium chromate and sodium nitrite solutions can be handled in iron and steel, and are in fact, effective corrosion inhibitors for ferrous alloys in water (see Chapter 35). On the other hand, if sufficiently alkaline, solutions of oxidizing salts may attack amphoteric metals (e.g., lead).

The hypochlorites are corrosive to copper, nickel, and lead but are easily handled by some plastics (e.g., PVC). They will not corrode steel to a significant extent unless they also contain free chlorine, as with certain proprietary formulations such as HTH (high-test hypochlorite). However, sodium hypochlorite is unstable in steel and can cause pitting. Alloy C-276 (N10276) is resistant but causes slow decomposition. Titanium is the preferred material for storage. EPDM is usefully resistant, while PVC may be used for temporary storage and feed piping in chlorination systems.

With the acidic salts, oxidizing cations such as ferric (Fe III) or cupric (Cu II) convey oxidizing characteristics to solutions which would otherwise act simply like dilute acids. Ferric sulfate or cupric phosphate are more aggressive than the corresponding calcium or magnesium salts, because their cations (being capable of reduction) act as cathodic depolarizers.

Oxidizing halides, like ferric chloride or cupric chloride, are very corrosive and cause pitting and/or SCC in susceptible alloys, as well as general corrosion, IGA, and sometimes selective weld attack. At moderately elevated temperatures, even alloy C-276 (N21076) is of limited usefulness. For the more aggressive conditions, only tantalum, glass, fluorinated plastics, or carbonaceous materials will contain them.

It is interesting to note that ferric chloride hexahydrate ($FeCl_3 \cdot 6H_2O$), in the presence of an excess of acid to suppress internal hydrolysis, has a vapor pressure about equal to that of mercury. It can therefore travel about a process as a vapor (where its presence may not be anticipated), leaving severe corrosion, pitting, or SCC in its wake, depending upon the material of construction. Steel may show localized corrosion of welds; copper and nickel alloys suffer general corrosion; and stainless steels suffer pitting, crevice corrosion, and/or SCC.

30

Hydrogen Phenomena

The little hydrogen atom is about one-third the size of the crystal cube in a metal lattice, and can easily diffuse through the structure of the metal. When the crystal lattice is saturated by atomic hydrogen, an adverse effect is observed on the properties of many metals and alloys. In the absence of oxygen and in the presence of a clean metal surface, the hydrogen molecule dissociates (which it is otherwise reluctant to do) and the hydrogen *atoms* are occluded in the metal structure. This can cause a reduction in strength and ductility. Nascent atomic hydrogen is also formed at cathodic sites in many electrochemical reactions.

Four separate phenomena are observed:

λ   Slow strain rate embrittlement (SSRE);

λ   Hydrogen blistering;

λ   Hydrogen-assisted cracking (HAC) and hydrogen-induced blister cracking (HIBC); and

λ   High-temperature effects.

The source of hydrogen may be corrosion (i.e., nascent hydrogen), electrochemical treatment (e.g., electroplating, cathodic protection) or high pressure, high temperature gaseous environments containing hydrogen (in which there is an equilibrium between molecular and atomic hydrogen).

30.1   Slow Strain-Rate Embrittlement

SSRE is a *temporary* loss of ductility when steel becomes charged with atomic hydrogen, as by pickling or electroplating (e.g., hydrogen plates out with cadmium during the operation). Ductility can be restored by heating for a short period of time at approximately 200°C (390°F), which drives out the hydrogen without permanent damage.

The outstanding feature of SSRE is that it is only detectable by *slow* strain, as in a tensile test. It does not show up in an impact test, for example. Field failures have occurred when storage tank roofs, saturated with hydrogen by corrosion in the presence of hydrogen sulfide, have been overpressured, resulting in a brittle fracture of the circumferential welds in the vapor space.

30.2 Hydrogen Blistering

In relatively soft steels, hydrogen blistering may be observed under mildly corrosive conditions in the presence of specific species (e.g., sulfides, selenides, arsenides, cyanides, or antimony compounds) which poison the dimerization of the nascent hydrogen formed. The nascent hydrogen is occluded in the lattice, where it dimerizes catalytically to molecular hydrogen at inclusions. Because the atomic hydrogen has weakened the lattice cohesion, the pressure generated by molecular hydrogen in the interstices of the metal causes blisters on the metal surface (sometimes also with splits or fissures). This tendency can be combatted to some extent by using steels of the same small grain size and cleanliness as are specified for improved nil ductility transition temperature (NDTT) in low temperature service.

30.3 Hydrogen-Assisted Cracking

This subject is discussed in Chapter 8.1.2.

30.4 High-Temperature Effects

Under partial pressures of hydrogen at elevated temperatures, the tendency for molecular hydrogen to dissociate is stronger and the mobile atomic hydrogen more active. Above approximately 400°C (750°F), for example, copper containing cuprous oxide inclusions will suffer localized reduction of the oxides *in situ* on exposure to hydrogen, generating high-pressure steam in the lattice, with attendant internal fissuring.

Of greater importance from the engineering standpoint is the *methanation* of steel. At temperatures above approximately 235°C

(425°F), atomic hydrogen will react with iron carbides (i.e., cementite) in steel, forming methane within the lattice and causing localized decarburization and fissuring. Additions of alloying elements like molybdenum and chromium tie up the carbon in more stable forms, permitting exposure to higher pressures and temperatures of hydrogen before methanation occurs. Note that traces of hydrogen sulfide work in the opposite direction, permitting methanation under less extreme conditions.

The parameters for successful use of ferrous alloys of increasing chromium content are defined by the "Nelson Curves," which are based on practical experience and observation and periodically updated by the American Petroleum Institute.[1] Simplified curves, for illustrative purposes, are given in Figure 30.1. A 0.5% molybdenum steel, originally used in the less extreme conditions before Cr-Mo steels are required, is no longer in common use. These curves indicate that at 13.7 MPa (2,000 psig) carbon steel

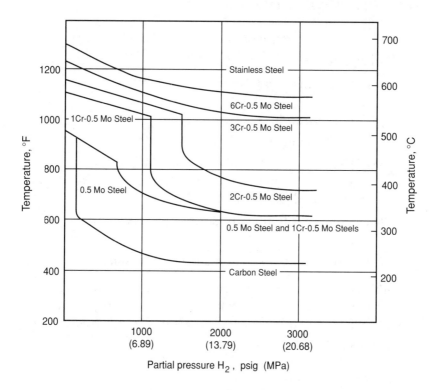

Figure 30.1 Simplified Nelson curves.

should be used up to approximately 230°C (450°F), 0.5% Mo Steel from approximately 230°C to 340°C (650°F), etc., and austenitic stainless steel above 600°C (1,100°F).

It is *very* important to note that atomic hydrogen goes right through the common metals. In an alloy-clad steel vessel (e.g., copper-, nickel-, or stainless-clad steel), the *back-up* steel must be alloyed to withstand the temperatures and partial pressure of hydrogen contained by the alloy layer on the process-side. Otherwise, the steel component of the pressure vessel is subject to methanation attack.

Only deoxidized copper, aluminum, zinc, cadmium, chromium, and silver (of the more common metals) are unaffected by exposure to hydrogen. They *are* permeable. Titanium, zirconium, tantalum, uranium, columbium, beryllium, and vanadium are all very reactive with hydrogen.

30.5 Practical Guidelines

In handling hydrogen at high pressures, the following precautions will help prevent equipment failures:

(1) Maintain low uniform hardness throughout the vessel;

(2) Require the highest quality welding, and check by thorough inspection;

(3) Use materials of maximum 552 MPa (80 ksi) yield strength (i.e., a hardness of approximately HRC 20 maximum);

(4) Reinspect frequently during the service life; and

(5) Always *vent* lined or multi-layer vessels.

Unlike other gases, hydrogen escaping under pressure gets *hotter* rather than colder. It can autoignite to form an almost invisible flame which is very hazardous to nearby structures as well as to personnel.

Reference

1. API 941, "Steels for Hydrogen Service at Elevated Temperatures and Pressures in Petroleum Refineries and Petrochemical Plants" (Washington, DC: API).

Suggested Resource Information

B. J. Moniz, W. I. Pollock, eds., *Process Industries Corrosion–The Theory and Practice* (Houston, TX: NACE International, 1986), p. 31.

31

High-Temperature Phenomena

It is somewhat difficult to define what is meant by "high temperature," because the different alloy systems go into specific modes of mechanical and corrosion behavior at different temperatures. Generally, a high temperature for any material is one at which it deforms under constant load and/or reacts with a low humidity gaseous environment (although molten salts or liquid metals also comprise high temperature environments).

The earliest demand for good high temperature materials arose in the power-generating industry, whose temperature requirements for boilers have risen steadily.

Year	Temp., °C (°F)	
1905	260	(500)
1926	400	(750)
1942	510	(950)
1956	620	(1,150)

At the present time, demands are also made by the petrochemical and nuclear industries, as for gas cracking in the 1,100°C (2,000°F) range, and by aerospace requirements.

31.1 Metal Behavior

There is a fundamental change in a metal's behavior when the temperature exceeds approximately 35% of its absolute melting range (or 40 to 60% for some nickel-based and cobalt-based alloys). Below that value, application of mechanical loads below the yield strength results in no permanent deformation when the load is removed. After the load is released, the metal item resumes its original dimensions, for all practical purposes. (It should be noted, however, that metallurgical changes can and do occur at relatively low temperatures; e.g., tempering, recrystallization, age-hardening, embrittlement, etc.)

Above some critical temperature, *creep* becomes a factor. Creep is time-dependent strain under constant load. The longer the load at some temperature is endured, the greater the permanent deformation (until rupture occurs). Creep is expressed as percent deformation per thousand hours. A typical creep curve is illustrated in Figure 31.1.

Design stresses for bolting and for unfired pressure vessels are 100% of the creep strength for 0.01% per 1,000 hours at temperature. ASME power boilers may also be designed at 60% of the 100,000-hour rupture strength (Figure 31.2).

31.2 Internal Stability

Temperature alone, without regard for the environment, can effect metallurgical changes in an alloy. In addition to the changes already mentioned, one mayencounter *temper* or *blue embrittlement* in the 455 to 600°C (850 to 1,100°F) range and/or graphitization of steel. Sensitization can occur in austenitic stainless steels and formation of intermetallic compounds, such as sigma or chi phase, in specific alloy systems. Such internal changes may affect mechanical properties or corrosion resistance, or both.

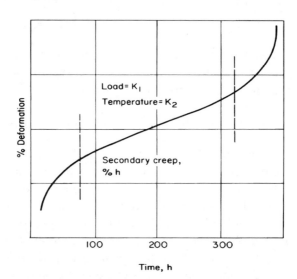

Figure 31.1 Time-deformation (creep) curve.

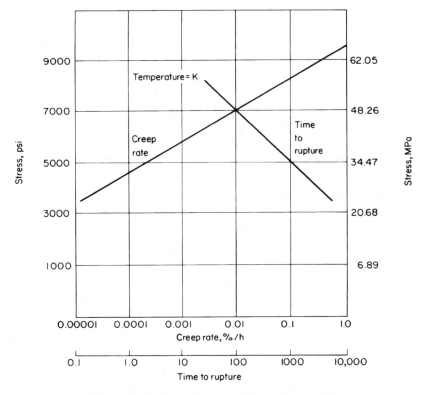

Figure 31.2 Creep Rate and Stress-Rupture Time.

31.3 Surface Stability

Depending on the nature and activity of the hot environment, an alloy may be subject to carburization, decarburization, or nitriding (sometimes severe, sometimes of a superficial nature).

31.4 High-Temperature Corrosion

All forms of attack at elevated temperatures in which the metal is converted to a corrosion product (solid, liquid, or vapor, depending on the metal/environment combination) are considered to be *oxidation*, in a manner analogous to aqueous corrosion. There is even electron transfer involved, and an "electrolyte" (i.e., the semiconductive layer of corrosion products). There is a definite migration of ions and electrons between the three phases (i.e., the metal, corrosion product layer, and environment). Unlike metals, electron transfer increases with temperature in a semi-conductor.

There is an ordering of metals at high temperature, analogous to the electromotive series. Gold will remain bright and unoxidized up to its melting point, while less noble metals oxidize more easily. The important thing to note is the *reversibility* of the reaction of metals with oxygen to form the oxide, depending on the specific metal, the temperature, and the nature of the environment.

For example, mercuric oxide rapidly releases oxygen at atmospheric pressure at approximately 500°C (930°F). On the other hand, for nickel oxide to release oxygen at 1,200°C (2,200°F) requires that the partial pressure of oxygen be reduced to below 10^{-7} atm (0.01 Pa). Otherwise, the nickel oxide will remain stable.

Just as the electromotive series does not apply when one departs from pure metals at standard conditions, diagrams relating the parameters for metal/metal oxide equilibria in different types of atmospheres (e.g., the Richardson diagram) do not apply to complex alloys. Also, the thermodynamics of a situation (which tell what *can* happen) do not reveal the kinetics of what *will* happen.

The rate at which a metal will oxidize depends upon how protective the oxide layer is. If it is wholly non-protective, the rate (i.e., usually weight gain) will be *linear* with time. If it is protective (and remains in place), the rate will be *parabolic* or *logarithmic*, diminishing with time. The structure of the oxide film also determines how easily metal ions can diffuse out (and gaseous species in).

In practice, cycling temperatures may spall off the surface oxidation products, leading to a *paralinear* rate (which just means that oxidation proceeds in a step-wise fashion). Changes in the nature of the environment, of course, can also remove or modify the surface products, as can mechanical effects.

31.4.1 Oxidation/Reduction

In hot air, oxygen, steam, carbon dioxide, etc., the environment will tend to oxidize a metal. With hydrogen, hydrogen-rich gases, or carbon monoxide, the environment is reducing, and tends to convert oxides back to the metallic state.

In *mixtures*, the ratio of carbon monoxide to carbon dioxide determines the carburization or decarburization conditions. The generalization extends to other combinations of oxidizing and reducing species, such as hydrogen and water vapor, nitric oxides and ammonia, and sulfurous oxides with hydrogen sulfide. An atmosphere may be reducing to one component such as nickel but oxidizing to another such as chromium or silicon, which further complicates the picture. The most reliable data are those determined experimentally.

Catastrophic oxidation may also occur, as when silicon oxide dissolves in nickel, or when molybdenum is vaporized as the oxide (e.g., from type 316 under insulation or deposits).

31.4.2 Sulfidation

Sulfidation is directly analogous to oxidation, but is aggravated by the lower melting points of many metallic sulfides and by the formation of low-melting eutectics (as in welding of sulfur-contaminated nickel alloys). Also, sulfide scales are less protective than oxide films.

31.4.3 Halogens and Hydrohalides

Because of the volatility of many metal halides, chlorine and hydrogen chloride are very aggressive at elevated temperatures. Fortunately, nickel alloys tend to be resistant. However, gold and silver are readily attacked by hot chlorine. While in hydrogen chloride, silver is attacked at approximately 230°C (450°F), gold is good to approximately 870°C (1,600°F).

31.4.4 Molten Salts

Many molten salts, alkaline or halide, attack metals at high temperature because they flux the surface films, continuously exposing fresh metal to attack. This is analogous to activation of passive metals, or solubilizing of films by complex metal ion formation, in aqueous corrosion.

31.4.5 Fuel Ash Corrosion

This is a special form of hot salt corrosion, in which the combination of sodium sulfate and chloride with vanadium compounds (as from certain types of crude oil fuels, for example) provides a very low-melting eutectic compound. This aggravates the fluxing, causing it to occur at lower temperatures than one would otherwise expect.

31.4.6 Molten Metals

Certain molten metals (e.g., mercury, lead, bismuth, etc.) are used as coolants in nuclear reactors. A sodium-potassium alloy (NaK) is also used. Such metals may dissolve a metal in the hotter sections by solubilization and then deposit it in the cooler sections (where solubility is lower), causing physical blockage.

Contamination of molten metals by their own oxides (e.g., NaO in Na or NaK) can also cause specific corrosion phenomena.

31.5 Effects of Alloying Elements

Alloying additions, such as chromium, aluminum and silicon, tend to form protective films which effectively limit the transport of reactive metals to the environment and of the environment inward to the metal. For practical purposes, these alloying elements enhance the formation of stable, low-volatility films, free of pores and adherent to the substrate. Nickel reinforces the effectiveness of chromium and also increases resistance to carburization, gas sulfidationm, and (up to approximately 80% nickel) attack by hot ammonia.

Chromium is the major alloying element in iron-, nickel- and cobalt-based alloys developed for high temperature resistance. Aluminum additions, while effective against oxidation, yield alloys which cannot be readily fabricated. Silicon is used as a minor, supplementary addition to increase the efficacy of chromium. Some rare earth metals also tend to improve the stability of the high temperature scale.

31.6 Heat-Resistant Alloys

This terminology usually refers to oxidation-resistant metals and alloys. However, it is necessary first to describe the temperature and oxidation limits for the carbon and low-alloy steels, which are in fact, used to at least moderately elevated temperatures.

31.6.1 Carbon Steel

Steel is not truly useful at temperatures above approximately 450°C (850°F), except in comparison with some non-ferrous alloys. Above this temperature, problems are encountered with high oxidation rates, temper embrittlement, loss of mechanical strength, and graphitization. Graphitization involves the decomposition of iron carbides, liberating free graphite, and is particular deleterious in the heat-affected zone of welds.

31.6.2 Molybdenum Steel

The addition of approximately 0.5% molybdenum extends the usefulness of steel to approximately 565°C (1,050°F) maximum. On exceeding this limit, however, the familiar problems with oxidation, loss of strength, and graphitization recur. Today, this type of steel, although long a favorite for steam piping, has been practically abandoned in favor of more highly alloyed albeit still low-alloy steels.

31.6.3 Alloy Steels

1.25%Cr-0.5%Mo and 2.25%Cr-0.5%Mo steels are used in refinery applications, particularly, because of their good high-temperature resistance to hydrogen and traces of hydrogen sulfide. They are also used in some high- temperature steam applications.

Higher alloyed steels (total alloy content more than 5%) include the 5%Cr-0.5%Mo + Si steel, which is useful to approximately 815°C (1,500°F), and the 7% and 9% chromium steels used for high-temperature, high-pressure hydrogen services.

31.6.4 Stainless Steels

Because of their chromium content, one would expect good high-temperature resistance for stainless steels under oxidizing conditions. However, certain limitations obtain.

(1) The nickel-free, straight chromium grades are useful to approximately 650 to 760°C (1,200 to 1,400°F), although there is a problem with sigma formation and attendant brittleness after extended service.

(2) The type 300 series austenitic stainless steels, depending upon the specific grade, are useful to as high as 870°C (1,600°F); e.g., types 347 and 316. Above that temperature, declining strength and increasing oxidation limit their capabilities.

(3) Higher-alloyed stainless steels, such as types 309 (UNS S30900), 310 (S31000), 25%Cr-12%Ni, 25%Cr-20%Ni, and 21%Cr-33%Ni, extend the usefulness of the austenitic grades in terms of improved resistance to oxidation and carburization, although adequate strength is not obtained until the high-carbon cast variants are employed. (See Chapter 31.6.6.)

31.6.5 Specialty Wrought Alloys

These include a variety of complex iron-, nickel-, and cobalt-based alloys such as:

(1) Alloy 802 (N08802), Cr-Fe-Ni plus W, Cb and Ti;

(2) Alloy 601 (N06601), Ni-Cr-Fe; and

(3) Stellite[†] 6 (R30006), Co-27%Cr-4%W.

31.6.6 Heat-Resistant Castings

Heat-resistant castings are employed where the analogous wrought materials would have insufficient strength at temperature. The "work-horse" of the petrochemical industry is HK-40 (UNS J94204, 25%Cr-20%Ni-0.35-0.45%C), used for ethylene pyrolysis furnace tubing. HP-50 (N08705, 26%Cr-36%Ni-0.35-0.75%C) and modifications thereof are becoming increasingly common in reformers and in ethylene pyrolysis furnaces. Other high-carbon, high-silicon castings such as HT (N08605, 16%Cr-35%Ni-0.35-0.75%C) are used in heat-treating furnaces. Several producers offer modified versions of the latter, strengthened by small, controlled additions of W, Cb (Nb), Si, etc.

There is some evidence that internal porosity in centrifugally cast tubing aggravates the tendency for carburization of the internal surface of cracking sets. Much HK tubing purchased today is bored and honed to improve its resistance. A number of castings of proprietary composition are being marked, based on allegedly superior resistance to carburization.

[†] Trade name

Suggested Resource Information

G. Y. Lai, *High Temperature Corrosion of Engineering Alloys* (Materials Park, OH: ASM International, 1990).

B. J. Moniz, W. I. Pollock, eds., *Process Industries Corrosion–The Theory and Practice* (Houston, TX: NACE International, 1986), p. 45.

32

Effects of Mercury

Mercury is a heavy metal (sp. gr. 13.6, approximately 1.7 times as heavy as steel), which is liquid above approximately –40°C (–40°F). Everyone is familiar with its use in thermometers and barometers. Historically, it was also used in manometers in plant for flow measurements by differential pressure across an orifice plate. Instruments containing mercury have previously been a potential source of process contamination. Fortunately, today's electronic instrumentation largely precludes this specific source.

Mercury has an appreciable vapor pressure even at room temperature; approximately 0.1 mm at 80°C (180°F). It will readily travel through piping and equipment as a vapor, contaminating the metal and causing specific effects.

The effects of mercury contamination on metals and alloys take three different forms:

(1) Amalgamation;

(2) Liquid environmental cracking (as LMC); and

(3) Increased general corrosion.

The effect encountered depends on the materials involved.

It should be remembered also that mercury is *poisonous* (specifically, it affects the nervous system—as in the old felt-manufacturing process, hence the expression "mad as a hatter"). An open container of mercury will significantly contaminate the air in a room.

32.1 Amalgamation

Among the common materials of construction, lead, tin, copper, and aluminum are very susceptible to amalgamation (i.e., the formation of a low-melting eutectic).

Aluminum alloys are very susceptible once the mercury penetrates the oxide film on the surface. Of course, mercury *ions* have the same effect, due to cementation. The resulting alloy of aluminum and mercury is subject to rapid attack even in such otherwise innocuous media as moist air or distilled water (presumably because no protective oxide film can be formed). Catastrophic corrosion has been encountered in aluminum storage tanks for chemically pure acetic acid, when contamination of the product was caused by mercury instrumentation.

Amalgamation of copper-nickel and nickel alloys is observed only at elevated temperatures, occurring rapidly at approximately 400°C (750°F) and only over an extended period of time at lower temperatures (see Chapter 32.2 concerning LMC).

32.2 Environmental Cracking

The metals which are subject to environmental cracking by mercury (either by the metal itself or by solutions of mercury salts) are normally copper, nickel-copper (alloy 400 [UNS N04400], etc.), nickel (N02200, etc.). Mercury can cause LMC of aluminum, as in liquified natural gas (LNG) heat exchangers. Ordinary iron and steel are unaffected. Both straight chromium and type 300 series stainless steels have been reportedly affected, but probably not by mercury alone. (Environmental cracking of austenitic grades is reported in mercuric chloride solutions, but it is not clear whether the phenomenon is influenced by the mercuric ion.)

The mechanism of cracking in nonferrous alloys appears to be a selective amalgamation at the grain boundaries, followed by mechanical damage to the weakened material by residual or applied stress. It is interesting to note that residual tensile stress is a prerequisite for attack by mercury salts. A standard mercuric nitrate test will differentiate degrees of cold-work from annealed material in yellow brass, for example.

Because of the very severe damage and the catastrophic rate of failure, we distinguish this mode of attack from the gross amalgamation described above.

32.3 Accelerated Corrosion

Although there is no amalgamation effect known, reliable

investigators have reported definitely accelerated corrosion of type 304 in salt solutions to which as little as 0.5 ppm of mercury have been added.

SECTION V

Elements of Corrosion Control

33

Corrosion Control

As previously indicated, there are five basic approaches to corrosion control. These are:

(1) A change in the material of construction;

(2) A change in the nature of the environment;

(3) Placing a barrier film or coating between the material and the environment;

(4) Application of an electrochemical potential to the material; and

(5) A change in the equipment design.

Sometimes, these concepts overlap to a certain extent, as we will see in the following detailed discussions.

33.1 Change of Material

A change in material of construction is commonly required to meet changing conditions. The change may be total or partial.

33.1.1 Total Changes

A total change in materials of construction is a rather obvious step in many circumstances, as where the anticipated life for steel has not been obtained in atmospheric or water exposures. Copper, nickel, or stainless alloys might be substituted to obtain a more economical life.

It must be remembered that, from a standpoint of economics, it is usually necessary to accept some finite life for any specific material and that there is an equivalent uniform annual cost, as discussed in Chapter 34.

33.1.2 Partial Replacement

At other times, corrosion may be economically controlled by changing only the *surface* in contact with the environment, or perhaps only a part of the equipment. The former situation is exemplified by the use of a substantial heavy-duty surface (as distinct from thin coatings).

Surface Barriers

These would include metallic linings, claddings, and weld overlays, as well as the organic and inorganic barriers, all of which are discussed in more detail in Chapter 37. The thinner platings, paints, and coatings are not considered in this category.

Vessel Parts

Sometimes, all that is required is to change a part of a vessel, such as heat exchanger tubes, column trays, pump impellers, or valve trim, or install an additional component (such as an impingement plate or entrainment separator). Such changes should always be made with proper consideration of galvanic effects.

A classical example of partial replacement is the practice of *safe-ending* heat exchanger tubes. Where the inlet tube ends are subject to localized phenomena (e.g., SCC, inlet-end erosion), it is sometimes economical to weld a short length of corrosion-resistant tubing to the longer length of conventional material (e.g., a short length of alloy 800 [UNS N08800] to type 304L [S30403], or alloy 825 [N08825] to type 316L [S31603]). A 30- to 45-cm (12- to 18 in.) length of resistant material in the hotter, more susceptible section of a heat exchanger bundle can obviate the need for a total retubing. However, the orbital (i.e., circumferential) welding of the individual short sections to the longer tubes is quite expensive, and the economics of safe-ending vs total retubing must always be carefully evaluated.

33.2 Change of Environment

Within this category of corrosion control lie such measures as total change (e.g., from air- to water-cooling), neutralization of acidic or alkaline species, scavenging of corrosive species, inhibition, and even simple changes in temperature and/or pressure.

33.2.1 Total Changes

Total changes in environment, to be economically feasible, usually must be anticipated in the design stage. To change from water-cooled to air-cooled condensers, from seawater to fresh-water systems, or from once-through to recirculated cooling water systems usually requires extensive redesign.

33.2.2 Partial Changes

Partial changes in the environment are often feasible. Indoors, relative humidity can be controlled and corrosive agents excluded by filtering and air-conditioning operations. For immersion conditions, the chemistry of the solution may be changed. Some common approaches are discussed below.

Acidity and Alkalinity Control

In aqueous media, pH is commonly controlled by minute additions of acid or alkali to establish conditions substantially noncorrosive to the materials of interest. In this manner, one may sometimes alleviate corrosion of steel and many nonferrous alloys or relieve the SCC propensity of austenitic stainless steels. As previously described, a high pH is conducive to the development of protective films for some alloy and solution combinations, such as calcareous deposits on steel in certain waters. Dilute acids, as in waste water or aqueous process streams, may be brought into some degree of control by neutralizing to a pH of 6 to 8, in many cases.

In organic liquids, both acid and alkaline constituents may function only in their proportion to contained or dissolved water (as opposed to their nominal weight per cent). In such cases, corrosion may be greater than anticipated from the percentage value, and partial or complete neutralization of either contained caustic or contained acid may be necessary.

Scavenging

Dissolved oxygen (DO) is frequently scavenged from water through treatment with sodium or ammonium sulfite, or hydrazine. As described in Chapter 19, catalyzed sulfite or hydrazine is available to hasten the speed of reaction.

Organic peroxides may be scavenged by reaction with sodium

nitrite in some hydrocarbon applications, reducing the oxidizing capacity of the process environment.

Not so widely known is the capability of propylene oxide and butylene oxide for scavenging traces of HCl from wet organic media, converting the HCl from the corrosive form to the noncorrosive propylene chlorhydrin.

Inhibitors

An inhibitor is a substance that reduces corrosion rates when relatively small amounts are added to the environment. Strictly speaking, an inhibitor should not be a major reactant, as in adjusting pH or scavenging oxygen. Inhibitors are discussed in detail in Chapter 35.

33.3   Anticorrosion Barriers

These comprise some of the metallic and organic or inorganic barriers discussed in Chapter 37, as well as the paints and coatings covered in Chapter 36. Usually, these are less than 1/8 in. (3.2 mm) thick, as distinct from heavy-duty applications discussed under changes in materials of construction in Chapter 33.1 above.

33.4   Electrochemical Techniques

Cathodic protection (CP) and anodic protection (AP) have been mentioned in previous discussion. A fuller treatment is given in Chapter 38.

There are growing applications of techniques for holding process equipment at some predetermined potential for the alleviation of pitting, crevice corrosion, or environmental cracking. These are really field applications of the laboratory potentiostat, and are different from conventional CP or AP techniques in that current is applied in either direction, as required by the metal's response to process changes.

Also in this category lies the use of galvanic couples for the detection of upset (corrosive) conditions, as described in Chapter 9.

33.5   Design Considerations

Design considerations in corrosion control vary from the

simple corrosion allowance against general attack to highly specific considerations for particular types of equipment.

33.5. 1 Corrosion Allowance

In the happy circumstance that only general (i.e., uniform) corrosion is to be expected, at a constant or decelerating rate, a useful life can be achieved through the use of a corrosion allowance. This comprises an extra thickness of the wall of the vessel, pipe or tubing which, when corroded away will still leave sufficient wall thickness to meet the mechanical design requirements at temperature and pressure. Such an approach is practical if a reasonable corrosion allowance of 3 to 6 mm (0.13 to 0.25 in.) will suffice, and metallic contamination is not a consideration.

33.5.2 Velocity Control

Excessive velocity may result in erosion-corrosion or, in certain cases cavitation.

Piping is traditionally limited to a liquid velocity of approximately 1.2 to 1.8 m/s (4 to 6 ft/s) in process applications. Heat exchanger tubing limits vary from as low as 1 m/s (3 ft/s) to 8 m/s (25 ft/s), depending upon the material of construction and the specific service.

Special problems arise at areas of locally high velocity or impingement (e.g., opposite calandria [reboiler] return lines in columns, at ells and tees, in tube inlet areas, or where internal weld protrusions, deposits or debris cause localized turbulence). For equipment inherently exposed to such effects, such as orifice plates, throttling valves, or pump impellers, erosion-resistant materials of construction are required.

33.5.3 Crevices and Drainage

Areas containing crevices or having inadequate drainage (e.g, flanges, threaded connections, internally protruding nozzles, or drain lines) are an invitation to pitting by oxygen concentration cells, localized accumulation of corrosive species, etc.

Crevices should be eliminated (e.g., by tube-to-tubesheet welding) or minimized, insofar as practical, and provision made for adequate drainage. Low "legs" in piping are to be avoided, as they may retain water or other corrosive or potentially corrosive materials, including sediment and deposits.

33.5.4 Stress Relief

There are two basic types of stress relief; thermal and mechanical. The former is more often employed against environmental cracking phenomena, the latter against mechanical fatigue. However, both are applicable as corrosion control measures against environmental cracking.

Thermal Stress Relief

Fabricated equipment is thermally stress-relieved as a palliative measure against corrosion fatigue or environmental cracking. It is also used to minimize stress from the standpoint of mechanical considerations or NDTT problems.

Thermal stress relief consists of heating the item of equipment to a temperature appropriate for the alloy involved, holding it at temperature for a specific period of time, usually for 1 h per in. (25 mm) of thickness and not less than 2 h, and then *slowly* cooling to room temperature (e.g., in the furnace or in still air).

Mechanical Stress Relief

Mechanical stress relief consists of shot-peening the surface of an item of equipment with an appropriate type of shot (e.g., steel, stainless steel, glass beads) to put that surface in a state of compressive stress. As previously explained, a surface in compressive (rather than tensile) stress is more resistant to mechanical fatigue, corrosion fatigue, or environmental cracking.

The shot-peening must be done under rigorously controlled conditions, using an *Almen strip*, which measures quantitatively the residual applied compressive stress (by the radius of curvature upwards resulting from the peening operation).

It is evident that mechanical stress relief will be effective only for so long as the surface under compressive stress is undisturbed by significant general corrosion.

33.6   Special Features In Heat Exchanger Design

Tube-and-shell heat exchangers present three problems related to their design and fabrication. These are: roll-leaks; tube defects (which can cause cross-contamination of one side of the exchanger or the other, possibly initiating corrosion problems which would not otherwise occur); and environmental cracking of tubes (which poses similar problems).

33.6.1 Tube-to-Tubesheet Joints

There are several potential problems associated with the conventional tube-and-shell heat exchanger fabrication.

Roll-Leaks

Roll-leaks in the tube-to-tubesheet mechanical joint may arise from stresses imposed when the tube material has a higher coefficient of thermal expansion than the shell (e.g., austenitic stainless steel bundles in carbon steel shells). Unless adequate provision is made to take up the stress (usually by means of a suitable expansion joint in the shell for heat exchangers of double-fixed tubesheet design), the tube ends will eventually be forced out of their rolled joints by the forces developed by the differential thermal expansion.

However, the problem can also arise when hard tubes are rolled into tubesheets of softer material (e.g., titanium tubes in a bronze tubesheet; superferritic stainless tubes in an type 304 tubesheet). There may also be severe deformation of the tubesheet ligaments in such cases.

Roll-leaks may also develop with soft tubes in harder tubesheets, but usually only under excessive thermal loads or cycles or as the result of multiple retubings of existing tubesheets.

Welded Joints

Welding the ends of the tubes to the tubesheet is not a substitute for adequate design against differential thermal expansion. However, both seal welds and strength welds have their particular uses.

A seal-weld, in which the tube-end is merely fused to the face of the tubesheet, may constitute insurance against mixing of the two streams in case the mechanically rolled joint should loosen. It is also employed where tubes are rolled into a clad tubesheet (to prevent ingress of the tubeside environment to the nonresistant substrate material) or where conditions for crevice corrosion of the rolled joint from the tubeside are severe. However, it is difficult to obtain a pore-free seal-weld, because of air trapped in the rolled joint (which tends to escape through the molten weldment).

A strength-weld of proper design is a more reliable joint, is just as easy or easier to make, and permits subsequent thermal stress-relief of the entire tube-bundle, if required (see further below). A good strength weld requires following proper procedures in proper sequence. One good procedure (see Figure 33.1), is as follows:

(1) "Drift" the tube into position in the tubesheet, using a tapered pin. The tube-end should protrude above the tubesheet for a distance equal to approximately 1 to 1½ times the tubewall thickness. *Do not roll;*

(2) Weld the tube-end to the tubesheet face, either with an autogenous weld or with filler metal (at the discretion of the welder);

(3) Apply an air-test from the shell-side; repair any leaking welds; and

(4) Complete the tube-rolling operation, starting approximately ¼ in. (6 mm) *behind* the weld joint.

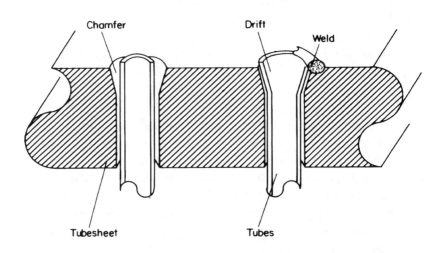

Figure 33.1 Strength-welded tube-tubesheet joint.

Since the strength now lies in the *welded* joint rather than in the rolled joint, the bundle can either be used in this condition *or* thermally stress-relieved, if desired. A number of other strength-welding procedures may be found in the literature.

33.6.2 Quality Control of Tubing

All tube-and-shell heat exchangers have potential problems as regards the integrity of the tubing. The quality of the tubing must be preestablished to ensure against cross-contamination at through-wall defects. In a gas-to-gas exchanger, a minor tube-wall leak may be of no consequence. However, in equipment exchanging heat between steam or water and a process stream, the consequences of a minor leak vary from minor to disastrous. Even a minor leak in either direction may have serious consequences in terms of either process contamination or corrosion or of environmental pollution.

Both seamless and welded tubes may be good, bad, or indifferent, depending on the care exercised during their manufacture and inspection.

Although the uninitiated tend to think of seamless tubes as a superior product, they can and do contain flaws and defects. If the extrusion process is done at too high or too low a temperature, the product can contain "extrusion tears." Lubricant or foreign matter may be retained on or in the tube wall. Stringers in the original billet may be carried through to form "line-capping," or lap defects may be formed by "chatter" during the extrusion process. Such tears or crack-like laps can form either through-wall defects or areas capable of triggering localized corrosion or cracking.

Welded tubes are made from strip or skelp, which has already been inspected for flaws. When the skelp is hot-formed and the mating edges longitudinally welded by automatic inert-gas shielded arc welding (without filler metal), a tubular product of superior concentricity and dimensional control is produced. Unfortunately, there is always concern about the integrity of that continuous longitudinal weld in the as-welded condition.

However, when a welded tube is cold-drawn to about 15% reduction in wall or 35% in area, prior to the final solution-anneal, the energy introduced by the cold-work permits the weld to recrystallize, forming small grains with a wrought structure similar to that of the parent metal. This severe cold-work also stretches and tests every inch of the weld. Such a produce is known as a *full-finished* tube and is very desirable for some critical corrosion services (e.g., hot organic acids).

For most services, aside from those highly conducive to localized corrosion (i.e., where minor defects might trigger cracking or pitting), it makes no difference whether the tubes are seamless, as-welded, bead-worked (i.e., *locally* cold-worked only on and along the longitudinal weld), or full-finished. However, *if* full integrity is required, then adequate inspection is required.

For *maximum* quality assurance, tubing should be subjected to non-destructive electronic inspection (i.e., ultrasonic or eddy-current, as appropriate), hydrostatic testing to 150% of the highest working pressure (tube- or shell-side), and also tested by air-under-water at 125 psig internal pressure.

33.6.3 Stress Corrosion Cracking

Because SCC of heat exchangers is a perennial problem, it seems worthwhile to emphasize certain aspects of design and operation.

Venting

Concentration of water-borne salts under the top tubesheet is a problem peculiar to vertical condensers having water on the shell (a design often favored to optimize condensation and subcooling). SCC occurs most often in the rolled area immediately below the tubesheet, especially with type 300 series stainless steel tubes.

To alleviate this problem, the top tubesheet should be vented at three or four locations around its periphery, so that there is no vapor space, and the face of the tubesheet is thereby continuously flushed and cooled by flowing water. The vents must be kept open and operable at all times. Note that even nonferrous tubes such as copper can fail by pitting or corrosion in the vapor space of a vertical condenser, unless tubesheet vents are provided.

Tube-Wall Temperature

It is desirable to design condensers so that the tube-wall temperature does not exceed the temperature at which calcareous deposits are formed by the cooling water. Usually, a temperature of 60°C (140°F) maximum is suggested, although 50°C (122°F) may be preferable for some hard waters of high chloride content.

Safe-Ending

Where the design is already fixed (as in retubing for maintenance purposes), or where the exchanger would be unduly large and expensive if designed for the suggested maximum tube-wall temperatures, the tubes may be safe-ended, as previously described.

Stress-Relief

Although stress-relieved *tubes* reportedly have been used, for example, for U-bundles in refinery exchangers, it must be remembered that additional stresses are imposed in assembly and operation (especially from the tube-rolling). To stress-relieve an entire strength-welded tube-bundle is a more reliable approach, as described above.

Suggested Resource Information

C. P. Dillon, D. R. McIntyre, *Guidelines for Preventing Stress Corrosion Cracking in the Chemical Process Industries*, MTI Publication No. 15 (Houston, TX: NACE International, 1985).

R. J. Landrum, *Fundamentals of Designing for Corrosion Control–A Corrosion Aid for Engineers* (Houston, TX: NACE International, 1989).

34

Introduction to
Economic Comparisons

Any recommendation for corrosion control involves asking management for money for cost reduction purposes. Although the technical basis for such a recommendation is presupposed, such requests must compete with others for capital investment or operating funds for other purposes and must stand on their own merits. The corrosion control recommendation(s) must be economically justified, and expressed in language that management can understand. Historically, a number of concepts from the discipline known as *engineering economy* have been employed. It is helpful to understand the basic concepts described below.

34.1   Return on Investment

Early on, the concept of *Return on Investment* (ROI), or its reciprocal concept of *Pay-Out Period*, was popular. ROI requires that a proposed saving to be effected by diminished *Annual Cost* (i.e., investment amount divided by years of service or *life*) must yield at least a stipulated percentage return on the *increase* in capital investment. Pay-Out Period (POP), the reciprocal, requires that the increase be paid back within a specified number of years.

$$ROI = \frac{\frac{I}{n} - \frac{I'}{n'}}{(I' - I)} \qquad (1)$$

$$\text{Pay-Out Period} = \frac{1}{ROI} \qquad (2)$$

where I and I´ are the old and new (increased) investments and n and n' are the respective lives.

Because I/n is not a true *Annual Cost* (see below) and because this calculation does not permit comparison of capital outlays versus expense items, another concept must be established if we are to have an adequately sophisticated approach.

34.2 Compound Interest

Compound interest has been called "the eighth wonder of the world" because, unlike simple interest, the *interest* earns interest. That is, interest is computed on the original *principal* together with its accrued interest. The principal is usually designated *present worth* (PW) and,

$$PW \, (1+i)^n \; = \; FW \qquad\qquad (3)$$

where FW is *future worth*, i is the interest rate, and n is the number of periods over which the principal is compounded.

34.3 Annual Cost or Savings

Another way to arrive at some FW is to put money aside at the end of each year, also at compound interest. It happens that end-of-the-year savings accrue according to the following formula:

$$FW = \frac{A \, (1 + i)^n - 1}{i} \qquad\qquad (4)$$

where A is the annual cost.

34.4 Present Worth and Annual Cost

Since things equal to the same thing are equal to each other, we have demonstrated that there is an *equivalent uniform annual cost* (A) for every present worth and future worth:

$$A = \frac{PW \, (1 + i)^n \times i}{(1 + i)^n - 1} \qquad\qquad (5)$$

or

$$A = PW \times i \times F_n \qquad\qquad (6)$$

where F is a variant from a "Sinking Fund" equation:

$$F = \frac{(1 + i)^n}{(1 + i)^n - 1} \qquad (7)$$

Note that $1,000 deposited in a savings account will draw interest over a period of years to attain some future worth. At 10% interest for 3 years:

$$\$1,000 \times (1.10)^3 = \$1,331 \qquad (8)$$

Another way to calculate FW is to set aside an annual saving, when:

$$\$402 \times \frac{(1.10)^3 - 1}{0.10} = \$1,331 \qquad (9)$$

and

$$\$402 = \$1,000 \,(0.10)\, F_3 \qquad (10)$$

where

$$F_3 = \frac{(1.10)^3}{(1.10)^3 - 1} = 4.02 \qquad (11)$$

These relationships demonstrate that, for the postulated conditions (i.e., money worth 10% over a 3-year period), a $1,000 PW, a $1,331 FW, and a $402 A (annual savings *or* cost) are mathematically equal.

34.5 Discounting

So far, we have been concerned with *now*, the present worth and its equivalent uniform annual savings (or cost). However, from the compound interest formula, it follows that if

$$PW \times (1 + i)^n = FW \qquad (12)$$

then

$$PW = \frac{FW}{(1 + i)^n} \qquad (13)$$

That is, any future worth can be *discounted* to its present worth. If money is worth 10% for a 3-year period, a *future* legacy of $1,331 has a present worth of $1,000.

This technique of discounting future worths is used to reduce them to equivalent present worths in order to sum them and then calculate the value of A. This permits us to compare all possible present or future income or expenditures on a directly comparable basis.

34.6 Depreciation

The Internal Revenue Service permits one to "write off" or *depreciate* large capital expenditures over a period of time (either the actual life of the equipment or some earlier established duration, e.g., 11 years for process equipment). Depreciation is defined as the lessening of value of an asset with the passage of time.

Depreciation may be physical, in the sense of deteriorating and becoming dysfuntional by reason of corrosion, etc. or functional, as by obsolescence, inadequacy, or inability to meet the demands placed upon the equipment.

There are a number of methods for depreciation, two of the most common depreciation schedules being *Straight-line* and *Sum-of-the Year's Digits*. In straight-line depreciation, the cost of an item expected to last three years is written off in thirds (i.e., one third for each year of the life, n). In sum-of-digit (SOD), the cost is written off in increments based on the fractions, the denominator of which is the SOD. For three years, the write-off fractions would be 3/6, 2/6, and 1/6, since 1+2+3=6. In general, the denominator is equal to $(n^2 + n)/2$ (i.e., for an 11-year write-off, the denominator would be $(122 + 11/2)$ or 66; the fractions 11/66, 10/66,..., 1/66).

The IRS permits a *tax credit* for each year's depreciation, which credit is equal to the depreciation times the tax rate. If we depreciate $1,000 capital investment over a three-year period by SL, with a 48% tax rate, the annual credit would be $1,000/3 × 0.48 or about $160 per year.

It should be emphasized that tax laws change from time to time, changing the permissibility and advantages of different depreciation schedules.

34.7 Discounted Cash Flow and PWAT

Discounted cash flow (DCF) is a technique used to arrive at *present worth after taxes* (PWAT). By setting up a capital investment for an anticipated duration, depreciating it on some permissible schedule, and taking the tax credit permitted, we establish first the cash flow itself. For example, a $1,000 investment having a three-year life, depreciated by SL, has the following cash flow:

Years

	1st	2nd	3rd
Investment	($1000)	(0)	(0)
Depreciation	$333	$333	$333
Tax Credit	$160	$160	$160
Cash Flow	($840)	$160	$160

These future values at the end of each year can be discounted by dividing them by $(1 + i)^n$ to get their present worth, which, because of the tax credit, are summed to PWAT. The DCF then becomes:

Years

	1st	2nd	3rd
DCF	($840)	$160	$160
	1.10	$(1.10)^2$	$(1.10)^3$
PWAT	($764) +	$132 +	$120
Total	($512)		

(*Note*: The PWAT of the capital investment is reduced by the PWAT of the future tax credits.)

Table 34.1 gives a list of factors under d_n, calculated from a standard accounting procedure which has been used in the chemical industry, based on a 48% tax rate and money worth 10%. Such factors allow us to reduce present worth to PWAT by multiplying by the factor, d, which is the precalculated present of a dollar in this procedure at year n. It follows that:

$$A = PW \times d_n \times r \times F_n \qquad (14)$$

Table 34.1

Economic Calculation Factors[A]

Years, n	$(1 + r)^n$	d_n	F_n
1	1.100	.520	11.000
2	1.210	.556	5.762
3	1.331	.583	4.021
4	1.464	.603	3.155
5	1.611	.617	2.637
6	1.771	.627	2.297
7	1.948	.633	2.055
8	2.145	.636	1.873
9	2.358	.639	1.736
10	2.595	.639	1.627
11	2.855	.640	1.539
12	3.14	.64	1.467
13	3.46	.64	1.407
14	3.80	.64	1.357
15	4.18	.64	1.314
20	6.72	.64	1.175
25	10.82	.64	1.102
30	17.0	.64	1.061

[A] Sum-of-Digit Depreciation (11-year schedule: balance in year of failure); maximum write-off 11 years (process equipment); $r = 10\%$; Tax Rate = 48%.

where r (rather than i) is the interest rate of return after taxes:

$$r = i\,(1 - t) \qquad (15)$$

with t the tax rate as a decimal.

Of course, the computer spreadsheet has vastly simplified the tabulation and calculation of discounted cash flow data.

34.8 PWAT and Annual Costs

We can now sum a whole series of financial transactions (income or outgo, plus or minus), regardless of when they occur in time. They can all be expressed in terms of PWAT and, ultimately, as equivalent uniform annual cost, A.

For example, we propose to spend \$X now and \$Y a year from now, and expect to receive \$Z ten years hence. Then

$$PWAT = -\$Xd_n - \frac{\$Yd_n}{(1+r)^1} + \frac{\$Zd_n}{(1+r)^{10}} \quad (16)$$

and

$$A = PWAT \times r \times F_n. \quad (17)$$

Note also that the Annual Costs can be individually calculated and summed.

$$A = -\$Xd_nF_n - \frac{\$Yd_nrF_n}{(1+r)^1} \cdots \quad (18)$$

34.9   Use of Annual Cost Calculations

The following sections discuss the general application of equivalent uniform annual costs, with specific illustrative examples.

34.9.1   Formulation

Annual cost calculations permit a direct comparison between alternatives of different duration and different accounting (e.g., capital items vs maintenance expenses), even when they occur at different times. (Similar present worth techniques are used for Life-Cycle Costs for alternatives with a single time period.)

The Annual Cost of any cash disbursement or income at any time can easily be defined if the following questions are asked:

(1) How much does it cost or pay (i.e., x)?

(2) When does it occur (i.e., $1/(1+r)n'$, where n' is the number of the year of occurrence hence)?

(3) What depreciation factor applies (i.e., d_1 for expense, d_n for capital)?

(4) What is the expected life?

Then

$$A = \pm\frac{x\, d_n\, r\, F_n}{(1+r)^{n'}} \quad (19)$$

where n' is the year of occurrence from now, n being the life.

34.9.2 Examples

A stainless steel pump costs $10,000 and lasts 2 years. The production loss associated with the shutdown costs $10,000 also. What is the annual cost?

$$A = - \ \$10,000 \ d_2 \ r \ F - \frac{\$10,000 \ d_1 \ r \ F_2}{(1 + r)^2}$$

$$A = - \ \$10,000(.556).1(5.762) - \frac{\$10,000(.52).1(5.762)}{1.21}$$

$$A = - \ \$3203 - \$2476 \ = - \ \$5679$$

Under these conditions, how much can we afford to pay for an alloy pump which would last five years? (Note that this only *delays* the shutdown cost from the second to the fifth year.) Since we must not spend more than we have already, we equate:

$$- \ \$5,679 = - x d_5 \ r f_5 - \frac{\$10,000 \ d_1 \ r \ F_2}{(1 + r)^2}$$

$$- \ \$5,679 \ = \ - x \ (0.1627) - \frac{\$1,371}{1.611}$$

$$- \ \$4,828 \ = - x \ (0.1627)$$

$$x \ = \ - \ \$29,675$$

Note that $29,675 is the *most* money that can be spent for a more resistant pump under the conditions postulated. If that much was spent, there would be no advantage. If it cost $20,000 only, A = $4,105 with an annual saving of $1,574.

34.9.3 Modern Developments

A complete discussion of this subject, including different accounting procedures and concepts, can to be found in NACE Standard RP0272.[1] The revision in the 1990s will include a program for the personal computer, tentatively entitled "Econocor," which permits rapid and facile economic comparisons to be made.

Reference

1. NACE Standard RP0272, "Direct Calculations of Economic Appraisals of Corrosion Control Measures" (Houston, TX: NACE International).

35

Corrosion Inhibitors

Corrosion inhibitors are widely used in water, acid-gas scrubbing systems, paint, oil-field applications, etc. Their use in chemical processes is of much lesser import. Nevertheless, the chemical engineer in the process industries should have at least a nodding familiarity with the subject.

35.1 Definition

A corrosion inhibitor is a chemical or combination of substances that, when present in the proper concentration and forms in the environment, prevents or reduces corrosion (ASTM G 15).[1] Strictly speaking, the inhibitor slows down or prevents corrosion *without itself taking a significant part in the reaction.* (Note that this definition should exclude treatment which radically changes the nature of the environment, such as deaeration, neutralization, or use of substantial additives, e.g., liming of soil.)

35.2 Electrochemistry

True inhibitors (as opposed to materials which decrease corrosion by reactivity with the environment, such as scavengers of dissolved oxygen (DO) or of HCl) function by adsorption on the metal surface, increasing polarization. A combination of adsorption of inhibitor and accumulation of corrosion products on the metal surface may produce a passive layer (e.g., chromates promote an invisible, protective oxide layer on steel in fresh water; small amounts of zinc cause precipitation of a protective hydroxide).

Corrosion inhibitors can affect either anodic or cathodic reactions or both. The classification is determined by whether the prior corrosion potential shifts in the positive (anodic) or negative (cathodic) direction. *Anodic inhibitors* are more dangerous because, added in insufficient amounts, they may actually *increase* general corrosion (functioning simply as cathodic depolarizers) or permit intensified attack (e.g., pitting) at sites devoid of inhibitor coverage. This is due to the larger effective cathode areas vs the small residual anodic areas inadequately covered by adsorption. With *cathodic inhibitors*, insufficient additions simply let localized corrosion at the anodes proceed without any acceleration of attack.

In most industrial applications in aqueous environments, the potential shifts are not usually sufficient to account for the entire inhibitive effect, indicating that both anodic and cathodic reactions are inhibited (i.e., the commercial inhibitors are a combination of anodic and cathodic inhibitors).

Successful use of inhibitors (especially anodic inhibitors) requires that the surfaces to be protected be boldly exposed so that they can adsorb inhibitor in adequate amounts, with the film readily replenished. Any crevices may then lead to locally high rates of attack. Cathodic inhibitors are effective against pitting and crevice corrosion, because the occluded areas are anodic sites and the current flow is usually determined by the area of the cathode.

35.3 Aqueous Systems

Economic considerations usually effectively preclude the use of inhibitors in once-through water systems. (One exception is the addition of ferrous sulfate to seawater to diminish erosion-corrosion of copper alloy condenser tubes.) Potable water is not particularly amenable to inhibition, because of toxicity or health considerations. However, because of the corrosivity of aerated water to steel and cast iron, many studies have been made. Complex polyphosphates and glassy silicates have sometimes been successfully used in very low concentrations (e.g., 1 to 2 ppm of sodium hexametaphosphate, the *threshold* treatment).

As indicated in Chapter 19, it is relatively easy and economical to adjust the water chemistry and inhibit open or closed recirculated cooling water systems.

Other considerations being equal, the corrosivity of water towards steel is proportional to the chloride content of the water (up to approximately 6,000 ppm chloride ion). Oxidizing inhibitors have traditionally been preferred, from the standpoint of efficiency, when the chlorides were above approximately 200 ppm. Chromates have been the popular choice from the standpoint of cost and effectiveness, as nitrates and nitrites not only are quite readily reduced but are a potential source of ammonia contamination, which is objectionable in many systems. On the other hand, chromates are objectionable today, from the environmental standpoint.

The amount of chromate required to inhibit a given chloride level in water can be reduced by judicious mixing with other inhibitors (e.g., polyphosphates, zinc, organic compounds). In some cases, such chromate-type inhibitors are still employed, the toxicological and ecological objections being met by treatment of the effluent prior to discharge.

There have been new non-chromate-type inhibitors developed (e.g., combinations of phosphates, polyphosphonates, and phosphate esters). Originally somewhat inadequate at chloride concentrations above perhaps 1,000 ppm, the newer formulations seem to be more effective, provided they are carefully controlled.

For a given water system, it is advisable to have the professional input from either a reputable water-treating firm or an independent consultant.

35.4 Refrigeration Brines

Solutions of approximately 25% calcium chloride are used in industrial processes to effect cooling below 0°C (32°F). These solutions are rendered noncorrosive by a combination of pH adjustment (e.g., 8.5 +) and inhibition with approximately 2,000 ppm chromate (as sodium chromate). Such solutions must be protected from exposure to air to prevent absorption of carbon dioxide, which will lower the pH and interfere with inhibition. If the pH drops below approximately 6.5, serious corrosion will occur. If caustic additions to maintain pH 8.5 to 9.0 are continued without excluding carbon dioxide, precipitation of calcium carbonate will occur. An inert atmosphere and good chemical control are required for these systems.

It is important not to alternate steam with calcium chloride or sodium chloride brines (or, for that matter, even low-chloride waters) in either steel equipment, in which serious corrosion will ensue, or austenitic stainless steel equipment, which is susceptible to serious pitting and/or SCC. The so-called "dimpled jacket" for vessels is also very susceptible to failure in chloride brines, because of the multiplicity of crevices involved in its spot-welded construction and the high localized stresses.

Aqueous solutions of methanol or ethylene glycol are also used for refrigerating "brines." These are less corrosive than the chloride solutions but nevertheless require treatment with borax (sodium tetraborate) to neutralize organic acids formed by oxidation, nitrites, and additions of such inhibitors as mercaptobenzothiazole (MBT) benzotriazole, or tolyltriazole.

35.5 Acids

Dilute solutions of hydrochloric acid (e.g., 5 to 15%) are used for pickling metal parts and in chemical cleaning of boilers and process equipment. These solutions are usually inhibited against corrosion of steel (*not* cast iron or stainless steel) with approximately 2,000 ppm of a proprietary inhibitor. The useful inhibitors include amines such as anilines and methyl ethyl pyridines, thiourea, sulfonated oils, acetylenic alcohols (e.g., propargyl alcohol), iodides, etc.

These types of inhibitors are also employed with dilute sulfuric and phosphoric acid solutions (preferably augmented with iodide additions, in the latter case).

The inhibitors for steel are not effective for cast iron, probably because of the galvanic influence of the free graphite. Also, stainless steels must not be exposed to hydrochloric acid cleaning solutions. The nonferrous alloys must be evaluated in terms of the specific acid and inhibitor combinations.

35.6 Acid-Gas Scrubbing Systems

The corrosion in acid-gas scrubbing systems, such as monoethanolamine (MEA), diethanolamine (DEA), and diglycolamine (DGA), needs to be appraised on the basis of individual systems. Some are corrosive, some not—depending upon mode of operation, degree of degradation, process contamination, etc.

Oil-field type inhibitors (see Chapter 35.7, "Nonaqueous Environments") are successfully used in some systems, while, in others, mixtures of inhibitors employed include sodium vanadate/

potassium antimony tartrate, etc. The oxidizing inhibitors are not useful when the gas stream contains hydrogen sulfide. They are used successfully in potassium carbonate/bicarbonate scrubbing systems as well as in ethanolamines.

35.7   Nonaqueous Environments

Inhibitors for petroleum products or oil-water mixtures usually feature an additive which is attracted to the metal surface via a polar group, and extends a water-repelling structure towards the bulk environment. These function in a manner analogous to the octadecylamine used as an inhibitor in steam condensate, as previously described.

Other inhibitors include cationic agents, like amines, and anionic agents, like sulfonates. The iodide ion apparently enhances the power of amines, such as decylamine, to function as corrosion inhibitors for steel.

Vapor-phase inhibitors (e.g., dicyclohexylammonium nitrite) are used in paper for wrapping finished machined parts as protection against atmospheric corrosion. They are also used in closing up equipment, as for dry lay-up. It should be emphasized that their throwing power is limited to short distances (e.g., a few inches), so they must be intimate contact with (or very close to) the surface to be protected.

Lastly, water itself is an effective inhibitor in some systems. The addition of 0.2% water to anhydrous ammonia inhibits SCC of steel.   Addition of 0.1 to 0.2% water inhibits the otherwise catastrophic corrosion of aluminum in anhydrous alcohols, and reduces corrosion of aluminum in glacial acetic acid. Carbon steel corrodes severely in 100% sulfuric acid, unless approximately 500 ppm water is added.   Although potentially dangerous unless the concentration is rigorously controlled, water is widely employed in such applications.

35.8   Other Considerations

The main considerations in selecting an inhibitor for a specific application are efficiency (because they usually only reduce, rather than prevent, corrosion), the economics of its use, and possible adverse effects. A particular corrosion inhibitor may have adverse effects on a process, even though otherwise effective, as by poisoning a catalyst or affecting final product quality. Particular components of a process equipment system may be incompatible with the inhibitor, even though it is protective of the major material of construction. It is important that valve trim, pump parts, process instrumentation, etc., be considered in making the final decision. It may be necessary to make significant changes in that area before implementing the corrosion inhibition program.

Reference

1. ASTM Standard G-15, "Definition of Terms Relating to Corrosion and Corrosion Testing" (Philadelphia, PA: ASTM).

Suggested Resource Information

B. J. Moniz, W. I. Pollock, eds., *Process Industries Corrosion–The Theory and Practice* (Houston, TX: NACE International, 1986), p. 603.

A. Raman, P. Labine, eds., *Reviews On Corrosion Inhibitor Science and Technology* (Houston, TX: NACE International, 1993).

36

Paints and Coatings

A tremendous amount of steel is used in both industrial and consumer applications, and much effort is expended on preventing its spontaneous rusting. Rust and corrosion can be a problem not only from the standpoint of wastage of metal but also from the aspect of product contamination (e.g., by internal corrosion of storage tanks). Machined parts can be rendered useless by even superficial rusting. Rust is also objectionable from the aesthetic viewpoint.

Temporary protection (as for machined parts; for steel in storage) may be achieved by means of special oils, greases, or wax-like coatings. More nearly permanent protection is effected by suitable paints or coatings, properly maintained.

In common parlance, the term *paint* is commonly used to indicate a system applied to the exterior, while *coatings* more often signify systems for internal lining or external spillage or other more severe conditions. However, NACE International terminology uses the word coatings in the broader sense to encompass all kinds of paints and coatings. The environments in which paints and coatings are employed includes the full range of aggressive acid, alkalis, and salts, usually coincident with moisture from air, steam, steam condensate, or process effluents.

Atmospheres are usually considered in three categories:

Type L (lightly corrosive)—industrial exteriors exposed to normal weathering (i.e., sunlight, wind, rain). There may also be present small amounts of fumes, vapors, or acid rain.

Type M (moderately corrosive)—involving chemical fall-out or fumes or occasional spillage of aggressive chemicals and/or steam or moisture condensation.

Type S (severely corrosive)—splash or spillage of aggressive chemicals and/or steam or moisture condensation. Usually cause general corrosion of unprotected steel at 3.0 to 6.0 mpy (0.08 to 0.15 mm/y).

Any one or combination of these categories may be present at a site.

36.1 Temporary Rust Preventatives

Ordinary oils and greases, such as those used for lubrication, are not themselves effective protection against rust. This is due to acidic components formed by the oxidation of hydrocarbons, and the fact that ordinary oils and greases cannot effectively displace water from the metal surface.

By the addition of buffering agents, corrosion inhibitors, and chemical additives which permit the oil to "wet" the metal surface (displacing moisture), common oils and greases are converted into effective rust preventatives.

These proprietary products, often called *slushing compounds*, vary in their effectiveness, depending on whether they are formulated and used for indoor or outdoor storage, for how long, and how aggressive the environment. (*Note:* An "indoor" environment may be very aggressive, as in the vicinity of acid fumes, as discussed in Chapter 21.)

The commercial rust preventatives vary in consistency from light machine oils through heavier oils, greases and waxes, to hard strippable films. Their characteristics vary depending on whether they are for indoor or outdoor use, and with or without a cover. Surface life can be from a few days to several years.

36.2 Paints and Coatings

Although everybody knows in a general way what paints and coatings are, they are not always defined in corrosion texts. For our purposes, they can be defined as follows.

36.2.1 Definition

Paints and coatings are a mixture of pigments with natural or synthetic resins (plus solvents, plasticizers, and extenders) which form a protective film by drying, oxidizing, or polymerizing. Currently, environmental restriction on volatile organic solvents, have pushed water-based formulations into greater prominence.

36.2.2 Types of Coatings

Paints and coatings may be based on either naturally occurring compounds, synthetic materials, or a mixture of the two. Traditionally, the natural type systems have been based on asphaltic, bitumenous materials, or on natural oils, such as those derived from rice, fish, etc. The latter group comprise the original "oleoresinous" paints, although the term has a much broader meaning today. These older systems are much more tolerant of poor surface preparation and contamination than are the modern synthetics.

Coating systems are categorized by generic type of binder or resin, and are grouped according to the curing or hardening mechanism inherent in that generic type. Although the resin or organic binder of the coating material is most influential in determining the resistances and properties of the paint, it is also true that the type and amount of pigments, solvents, and additives such as rheological aids will dramatically influence the application properties and protective capability of the applied film. Furthermore, hybridized systems can be formulated that are crosses between the categories. For example, the acrylic monomer or prepolymer can be incorporated with virtually any other generic type of resin to produce a product with properties that are a compromise between the acrylic and the original polymer. In many cases, this is advantageous, as in the mixing of vinyls and acrylics or heat-curing alkyds and acrylics. In other cases, such as with an epoxy, acrylic modification may be a detriment.

The modern synthetic coating systems are based on a variety of chemistries. Usually, they demand more sophisticated surface preparation and application than the natural-type systems. However, their improved performance more than compensates for the incremental cost of materials and labor.

Today, paints and coatings must be in compliance with volatile organic compound (VOC) restrictions and U.S. OSHA regulations. Federal VOC restrictions are 340 grams of organics per liter of paint solution, maximum. California has imposed limits of 275 g/L, and other states are considering similar tighter limits. Because many currently used paint systems cannot meet these limits, new technologies (discussed further below for specific systems) are available to address the VOC problem.

The various types of paints/coatings systems in current use include the following:

Alkyds. This nomenclature derives from the alcohols and acids used in the manufacture of the paint. Alkyds have largely replaced the original oil-based paints. They are faster drying and have better gloss retention, greater hardness, increased water resistance, good weathering characteristics, and good durability in mild exposures. Although they do not have good chemical or solvent resistance, they are the "work-horse" of much industrial painting.

Acrylics. These are systems based on resins formulated from acrylic acid and acrylate esters. They are used to produce a high-gloss finish on house siding, automobiles, and appliances. Because they also have good chemical and sunlight resistance, they are also being increasingly used in industry both for appearance and protection.

Direct-To-Metal (DTM) acrylics (or DTM primers). These are designed for new construction or maintenance applications in which color retention and gloss aesthetics are important. Used for both interior and exterior applications, they may be applied to steel, galvanized steel, iron, aluminum, and concrete without a primer. They provide resistance to corrosion, chemicals, UV, fumes, splash and spillage from selected acids, alkalis, alcohols, and hydrocarbon solvents. They offer the convenience of a water-based system with the performance qualities of conventional solvent-based systems.

Water-Borne Dry-Fall Acrylics. These are designed for high reflectance while permitting overspray to dry to a removable dust. They are used for high ceilings and walls in large warehousing facilities and may be applied to steel, galvanized, aluminum, masonry, or wood. They emit little odor during application and have a high flash-point conducive to safety.

Bitumenous. The bitumenous systems comprise asphaltic or coal tar-based resins, including both natural and combination natural/synthetic mixes. Coal-tar coatings have good water resistance (which can be improved by formulating with epoxies), but are not very good in weather and have poor resistance to sunlight. Bitumenous coatings are mostly used for underground protection.

Chlorhydrocarbons. Chlorinated hydrocarbon paints are quite similar to chlorinated rubber. Having good resistance to moisture and inorganic chemicals, they are particularly useful for application to masonry.

Chlorinated Rubber. These systems have good chemical and excellent water resistance, but are adversely affected by direct sunlight. They are good for splash-zone and immersion service, and are useful for coating concrete.

Epoxies. These are formulated from polyphenols and epichlorinhydrin, and require a catalyst to effect curing. The epoxies have good adhesion and are resistant to acids, alkalis, and many solvents. Good quality catalyzed epoxies are used for many severe atmospheric and water-immersion applications.

Water-Based Catalyzed Epoxies. These are designed for surface protection of interior high-maintenance areas. They provide a tile-like finish, with resistance to corrosion, chemicals, impact and abrasion, and may be applied to poured concrete, cinder block, dry wall, and gypsum surfaces. They can be used to upgrade conventional latex and alkyd-based coatings for improved resistance, without lifting or bleeding through the underlying coating.

Epoxy Mastics. These are designed for moderate to severe exposures. Compatible with old, intact paint films, they may be used on structural steel members, equipment and machinery housings. They may be applied to rusted and pitted steel surfaces where extensive surface preparation (as by blasting) is impractical. A 6 to 8 mil (75 to 200 μm) dry-film thickness per coat allows them to withstand aggressive abrasive cleaning and heavy wear and tear while offering resistance to acids, salts, chemicals, and moisture.

All-Weather Epoxies. These are designed for fast turnaround jobs at ambient temperatures, providing a dry-to-touch condition overnight, and amenability to cold weather application. They are used in corrosive environments where rapid temperature fluctuations may occur, providing resistance to abrasion, direct impact, chemicals, moisture, and bacterial effects for steel or masonry construction (e.g., offshore structures, refineries, storage tanks).

100-percent Solids Self-Leveling Epoxies. These are designed to provide moisture, chemical and cleaning solution resistance in production and high-traffic areas. They provide a high dry-film thickness for floor surfaces of smooth or mildly spalled concrete, as well as mild steel surfaces.

Epoxy-Esters. These systems are compounded with drying oils to provide an uncatalyzed, single-package paint. They have good resistance to moisture, but have sacrificed most of their chemical and UV resistance.

Latex. These water-based paints are formulated from vinyl or acrylic resins and have the advantage of lacking objectionable solvents. They are easily applied to masonry, as well as steel.

Oleoresinous. These are any combination of oils and resins, including oil-modified phenolics and modified epoxies, as well as alkyds. Because of their slow drying characteristics and limited chemical resistance they have only restricted application in modern industrial plants.

Phenolic. The outstanding phenolic systems are those which are *baked* at approximately 230°C (450°F) to provide a 3 to 5 mil (75 to 130 μm) coating of high chemical resistance. Although attacked by oxidants and by even dilute alkalis, they provide both corrosion and contamination protection in a wide variety of chemical and petroleum services.

Phenoxy. The phenoxy systems make excellent primers because they readily wet and adhere to steel. Because they lack good weathering characteristics, they are not used for finish coats. Their primary use is as zinc-rich organic primers.

Polyesters. Besides their use in reinforced plastics and high-build barriers, the polyesters and vinyl esters are available as both thin- and medium-build coatings.

Silicones. The silicone systems are quite expensive, being based on organic silicon compounds (which have silicon rather than carbon linkages in the structure). They are primarily used for high temperature service, where carbon-based coatings would oxidize. Silicone paints are frequently pigmented with powdered aluminum.

Urethanes. Several types of urethane formulations are defined in ASTM D 16.[1] These products provide a wide range of properties characterized by excellent abrasion resistance, good chemical resistance, and outstanding gloss retention. Proper selection of the appropriate formulation for the intended service is very important.

Vinyls. The copolymers of vinyl chloride and vinyl acetate are solvent-based paint systems. While they lack solvent resistance thereby, they are outstanding in marine environments. They also have good resistance to acids, alkalis, and oxidants. An excellent combination for industrial/marine exposures is a vinyl top-coat over an inorganic zinc primer.

Zinc, Inorganic. The "IZ" system is silicate-based, heavily pigmented (approximately 70%) with powdered zinc. When dried, the zinc content provides CP to the steel substrate at any holidays, i.e., through-film defects, or scratches. The IZ system is excellent in its own right, being resistant to high humidity atmospheres and salt spray (although it will not withstand acids or alkalis). However, it gives its best service when top-coated with another system (e.g., vinyl, epoxy, coal-tar, etc.). It should be noted that any sulfur contamination may make it difficult to top-coat. The choice between using IZ paint systems as opposed to hot-dipped galvanized steel is primarily an economic decision, largely based on the area per ton of steel.

36.2.3 Advantages and Limitations of Principal Coating Resins

Alkyds. *Advantages.* Good resistance to atmospheric weathering and moderate chemical fumes; not resistant to chemical splash and spillage. Long oil alkyds have good penetration but are slow drying; short oil alkyds are fast drying. Temperature resistant to 105°C (225°F). *Limitations.* Not chemically resistant; not suitable for application over alkaline surfaces, such as fresh concrete or for water immersion. *Comments.* Long oil alkyds make excellent primers for rusted and pitted steel and wooden surfaces. Corrosion resistance is adequate for mild chemical fumes that predominate in many industrial areas. Used as interior and exterior industrial and marine finishes.

Epoxy Ester. *Advantages.* Good weather resistance; chemical resistance better than alkyds and usually sufficient to resist normal atmospheric corrosive attack. *Limitations.* Generally the least resistant epoxy resin. Not resistant to strong chemical fumes, splash, or spillage. Temperature resistance: 105°C (225°F) in dry atmospheres. Not suitable for immersion service. *Comments.* A high-quality oil-base coating with good compatibility with most oil coating types. Easy to apply. Used widely for atmospheric resistance in chemical environments on structural steel, tank exteriors, etc.

Vinyls. *Advantages.* Insoluble in oil, greases, aliphatic hydrocarbons, and alcohols. Resistant to water and salt solutions. Not attacked at room temperature by inorganic acids and alkalis. Fire resistant; good abrasion resistance. *Limitations.* Strong polar solvents redissolve the vinyl. Initial adhesion poor. Relatively low thickness (1.5 to 2.0 mils [40 to 50 µm]) per cost. Some types will not adhere to bare steel without primer. Pinholes in dried film are more prevalent than in other coating types. *Comments.* Tough and flexible, low toxicity, tasteless, colorless, fire resistant. Use in potable water tanks and sanitary equipment; widely used industrial coating. May not comply with VOC regulations.

Chlorinated Rubbers. *Advantages.* Low moisture permeability and excellent resistance to water. Resistant to strong acids, alkalis, bleaches, soaps and detergents, mineral oils, mold, and mildew. Good abrasion resistance. *Limitations.* Redissolved in strong solvents. Degraded by heat (95°C or 200°F, dry; 60°C or 140°F, wet) and ultraviolet light, but can be stabilized to improve these properties. May be difficult to spray, especially in hot weather. *Comments.* Fire resistant, odorless, tasteless, and nontoxic. Quick drying and excellent adhesion to concrete and steel. Use in concrete and masonry paints, swimming pool coatings, industrial coatings, marine finishes.

Coal Tar Pitch. *Advantages.* Excellent water resistance (greater than all other types of coatings); good resistance to acids, alkalis, and mineral, animal, and vegetable oils. *Limitations.* Unless cross-linked with another resin, is thermoplastic and will flow at temperatures of 40°C (100°F) or less. Hardens and embrittles in cold weather. Black color only, will alligator and crack upon prolonged sunlight exposure, although still protective. *Comments.* Used as moisture resistant coatings in immersion and underground service. Widely used as pipeline exterior and interior coatings below grade. Pitch emulsions used as pavement sealers. Relatively inexpensive.

Polyamide-Cured Epoxies. *Advantages.* Superior to amine-cured epoxies for water resistance. Excellent adhesion, gloss, hardness impact, and abrasion resistance. More flexible and tougher than amine-cured epoxies. Temperature resistance: 105°C (225°F) dry; 65°C (150°F) wet. *Limitations.* Cross-linking does not occur below 5°C (40°F). Maximum resistances generally require seven day cure at 20°C (70°F). Slightly lower chemical resistance than amine-cured epoxies. *Comments.* Easier to apply and topcoat, more flexible, and better moisture resistance than amine-cured epoxies. Excellent adhesion over steel and concrete. A widely used industrial and marine maintenance coating. Some formulations can be applied to wet or underwater surfaces.

Coal Tar Epoxies. *Advantages.* Excellent resistance to saltwater and freshwater immersion. Very good acid and alkali resistance. Solvent resistance is good, although immersion in strong solvents may leach the coal tar. *Limitations.* Embrittles upon exposure to cold or ultraviolet light. Cold weather abrasion resistance is poor. Should be topcoated within 48 hours to avoid intercoat adhesion problems. Will not cure below 10°C (50°F). Black or dark colors only. Temperature resistance: 105°C (225°F) dry; 65°C (150°F) wet. *Comments.* Good water resistance. Thicknesses to 10.0 mil (0.25 mm) per coat. Can be applied to bare steel or concrete without a primer. Low cost per unit coverage.

Polyurethanes (aromatic or aliphatic). *Advantages.* Aliphatic urethanes are noted for their chemically excellent gloss, color, and ultraviolet light resistance. Properties vary widely, depending on the polyol coreactant. Generally, chemical and moisture resistances are similar to those of polyamide-cured epoxies, and abrasion resistance is usually excellent. *Limitations.* Because of the versatility of the isocyanate reaction, wide diversity exists in specific coating properties. Exposure to the isocyanate should be minimized to avoid sensitivity that may result in an asthmatic-like breathing condition upon continued exposure to humidity, which may result in gassing or bubbling of the coating in humid conditions. Aromatic urethanes may darken or yellow upon exposure to ultraviolet radiation. *Comments.* Aliphatic urethanes are widely used as glossy light fast topcoats on many exterior surfaces in corrosive environments. They are relatively expensive, but extremely durable. The isocyanate can be combined with other generic materials to enhance chemical, moisture, low-temperature, and abrasion resistance.

Asphalt Pitch. *Advantages.* Good water resistance and ultraviolet stability. Will not crack or degrade in sunlight. Non-toxic and suitable for exposure to food products. Resistant to mineral salts and alkalis to 30% concentration. *Limitations.* Black color only. Poor resistance to hydrocarbon solvents, oil, fats, and some organic solvents. Do not have the moisture resistance of coal tars. Can embrittle after prolonged exposure to dry environments or temperatures above 150°C (300°F), and can soften and flow at temperatures as low as 40°C (100°F). *Comments.* Often used as relatively inexpensive coating in atmospheric service, where coal tars cannot be used. Relatively inexpensive. Most common use is as a pavement sealer or roof coating.

Water Emulsion Latex. *Advantages.* Resistant to water, mild chemical fumes, and weathering. Good alkali resistance. Latexes are compatible with most generic coating types, either as an undercoat or topcoat. *Limitations.* Must be stored above freezing. Does not penetrate chalky surfaces. Exterior weather and chemical resistance not as good as solvent or oil-base coatings. Not suitable for immersion service. *Comments.* Ease of application and clean-up. No toxic solvents. Good concrete and masonry sealers because breathing film allows passage of water vapor. Used as interior and exterior coatings.

Acrylics. *Advantages.* Excellent light and ultraviolet stability, gloss, and color retention. Good chemical resistance and excellent atmospheric weathering resistance. Resistant to chemical fumes and occasional mild chemical splash and spillage. Minimal chalking, little if any darkening upon prolonged exposure to ultraviolet light. *Limitations.* Thermoplastic and water emulsion acrylics not suitable for any immersion service or any substantial acid or alkaline chemical exposure. Most acrylic coatings are used as topcoats in atmospheric service. Acrylic emulsions have limitations described under "Water Emulsion Latex." *Comments.* Used predominantly where light stability, gloss, and color retention are of primary importance. With cross-linking, greater chemical resistance can be achieved. Cross-linked acrylics are the most common automotive finish. Emulsion acrylics are often used as primers on concrete block and masonry surfaces.

Amine-Cured Epoxies. *Advantages.* Excellent resistance to alkalis, most organic and inorganic acids, water, and aqueous salt solutions. Solvent resistance and resistance to oxidizing agents are good as long as not continually wetted. Amine adducts have slightly less chemical and moisture resistance. *Limitations.* Harder and less flexible than other epoxies and intolerant of moisture during application. Coating will chalk on exposure to ultraviolet light. Strong solvents may lift coatings. Temperature resistance: 105°C (225°F) wet; 90°C (190°F) dry. Will not cure below 5°C (40°F); should be topcoated within 72 hours to avoid intercoat delamination. Maximum properties require curing time of about 7 days. *Comments.* Good chemical and weather resistance. Best chemical resistance of epoxy family. Excellent adhesion to steel and concrete. Widely used in maintenance coatings and tank linings.

Phenolics. *Advantages.* Greatest solvent resistance to all organic coatings described. Excellent resistance to aliphatic and aromatic hydrocarbons, alcohols, esters, ethers, ketones, and chlorinated solvents. Wet temperature resistance to 95°C (200°F). Odorless, tasteless, and non-toxic; suitable for food use. *Limitations.* Must be baked at a metal temperature ranging from 175 to 230°C (350 to 450°F). Coating must be applied in a thin film (approximately 1 mil [0.03 mm]) and partially baked between coats. Multiple thin coats are necessary to allow water from the condensation reaction to be removed. Cured coating is difficult to patch due to extreme solvent resistance. Poor resistance to alkalis and strong oxidants. *Comments.* A brown color results upon baking, which can be used to indicate the degree of cross-linking. Widely used as tank lining for alcohol storage and fermentation and other food products. Used for hot water immersion service. Can be modified with epoxies and other resins to enhance water, chemical, and heat resistance.

Organic Zinc-Rich. *Advantages.* Galvanic protection afforded by the zinc content, with chemical and moisture resistance similar to that of the organic binder. Should be topcoated in chemical environments with a pH outside the range 5 to 10. More tolerant of surface preparation and topcoating than inorganic zinc-rich coatings. *Limitations.* Generally have lower service performance than inorganic zinc-rich coatings, but ease of application and surface preparation tolerance make them increasingly popular. *Comments.* Widely used in Europe and the Far East, while inorganic zinc-rich coatings are most common in North America. Organic binder can be closely tailored to topcoats (for example, epoxy topcoats over epoxy zinc-rich coatings) for a more compatible system. Organic zinc-rich coatings are often used to repair galvanized or inorganic zinc-rich coatings.

Inorganic Zinc-Rich. *Advantages.* Provides excellent long-term protection against pitting in neutral and near-neutral atmospheric and some immersion services. Abrasion resistance is excellent, and dry heat resistance exceeds 370°C (700°F). Water-base inorganic silicates are available for confined spaces and VOC compliance. *Limitations.* Inorganic nature necessitates thorough blast-cleaning surface preparation and results in difficulty when topcoating with organic topcoats. Zinc dust is reactive outside the pH range of 5 to 10, and topcating is necessary in chemical fume environments. Somewhat difficult to apply; may mudcrack at thicknesses in excess of 5 mil (0.13 mm). *Comments.* Ethyl silicate zinc-rich coatings require atmospheric moisture to cure and are the most common type. Widely used as a primer on bridges, offshore structures, and steel in the building and chemical processing industries. Used as a weldable preconstruction primer in the automotive and shipbuilding industries. Use eliminates pitting corrosion.

36.2.4 Coating System Selection

The selection of the proper corrosion resistant coating system depends on a number of factors. This section will highlight the more important considerations.

(1)   Environmental Resistance—The coating system should be resistant to the chemical, temperature, and moisture conditions expected to be encountered in service.

(2)   Appearance—In high visibility areas (for example, water tanks, railroad cars, and appliances), color, gloss, and a pleasing appearance may be very important.

(3)   Safety—Toxic pigments and solvents may prohibit the use of some coatings; future removal and disposal problems should be considered.

Coatings with volatile explosive solvents may be dangerous in enclosed, poorly ventilated spaces.

(4)   Surface Preparation—Blast cleaning may be prohibited in some refineries or near electrical and hydraulic machinery.

(5)   Skill of the Labor Force—Certain coating systems require more application expertise than others. Generally, inorganic zincs, vinyl esters, polyesters, vinyls, and chlorinated rubbers are less application tolerant than some of the other coating systems.

(6)   Substrate to be Coated—Coating systems for aluminum, lead, copper, and so on, may be different from those for ferrous metals.

(7)   Available Equipment—The more resistant coating systems generally require spray application over a blast cleaned surface. For some polyesters and vinyl esters, multi-component spray equipment may be required.

(8)   Design Life—If a structure is intended to have a long service life, the use of a more resistant coating system may be justified than if a shorter service life is expected.

(9)   Cost—Generally speaking, the applied cost of a more resistant coating system is greater than the applied cost of a less resistant system.

(10) Accessibility for Future Repair—If future repair will be difficult or expensive, perhaps a longer lasting coating system should be specified.

(11) Consequences of Coating Failure—If a coating failure will be disastrous, such as in tanks holding highly corrosive materials or in nuclear power plants where peeling or disbonding coating may clog sumps and screens, a more resistant coating system should be selected.

Each of these factors will, to varying degrees, directly influence the choice of the coating system selected to protect a metal in a given environment. Once the choice is made, it is usually necessary to prepare appropriate specifications detailing the surfaces to be painted, and the surfaces to be protected from paint. In addition, the surface preparation and coating application details should be delineated. At a minimum, the thickness of each coat and total coating thickness should be specified. The surface profile of a blast cleaned surface, the interval between coats, and other factors (such as testing for holidays and pinholes and for adhesion) should also be specified. Although the preparation of a proper specification is beyond the scope of this article, a good specification is not only an important contractual document, but is also the means of conveying the owner's (or specifier's) intentions and instructions regarding the painting to be accomplished. Thus, the potential importance of a good specification cannot be overestimated.

36.2.5 Surface Preparation

Selection of the appropriate paint system presupposes consideration of the application conditions and material problems that may be encountered (e.g., temperatures to be expected, catalyst problems, compatibility with existing paints, viscosity problems, drying rates, contamination effects, overspray, flowout, thickness, moisture tolerance of the paint system, etc.). Certainly, the system must be durable—i.e., resistant to ultraviolet rays (UV), chalking, oxidation, loss of constituents after curing, thermal shock, moisture, chemicals, and loss of adhesion.

A variety of surface preparation methods are available, including:

λ   Hand Tool Cleaning;

λ   Power Tool Cleaning;

λ   Steam Cleaning;

λ   Water Blasting;

λ   Abrasive Blast Cleaning; and

λ   Solvent or Chemical Washing.

Legislation concerning worker health and environmental protection has had a marked impact on the coatings industry. All facets of coating technology are strongly influenced by legislative decree. The alkyds, vinyls, and epoxies currently in use are being reformulated, and within a few years (although the names will remain the same), the compositions of these coatings (and their potential protective capabilities) may be considerably different. Furthermore, surface preparation and application techniques in current use are also under considerable legislative pressure, and are expected to change considerably in the near future.

For example, although the practice of blast-cleaning a steel surface and applying an alkyd, vinyl, or epoxy coating system remains one of the best methods of corrosion protection for many steel surfaces. The alkyds, epoxies, and vinyls themselves have changed as a result of legislation, restricting the release of volatile organic compounds (solvents), and the use of toxic pigments.

The minimum in mechanical cleaning is derived from power wire-brushing or needle guns. Abrasive blasting (effected with sand, grit, or other materials, wet or dry) has been widely used to establish a suitable surface profile while cleaning the surface to promote both mechanical and ionic bonding. Blasting with high-pressure water has been used to avoid the fine particulate debris from abrasive blasting.

Most modern paint systems require good surface preparation, which is extremely important (some 80% of paint failures are due to inadequate preparation). Surface preparation can be a major part of the overall cost (usually 70 to 80%). Older methods such as brushing, scraping, and flame cleaning have largely given way to mechanical cleaning.

Surface preparation techniques have changed drastically; silica sand as a blast-cleaning abrasive has been banned in virtually all Western countries except the United States and Canada (although there is a strong movement to ban it in these countries as well). The containment and safe disposal of spent blast cleaning abrasives are required in many localities to prevent environmental damage caused by the leaching into water supplies of lead, chromate, and other toxic paint pigments removed during the course of blast cleaning. Therefore, although a paint layer over a properly cleaned surface still acts as a barrier against a corrosive environment in most cases, the components that constitute this barrier have changed considerably within the past ten years, and the means by which the surface is properly cleaned are rapidly changing at the present time.

Surfaces should be clean, dry, and dull. NACE International Standards are available, consisting of steel coupons which illustrate the degree of cleanliness (e.g., brush-off, commercial, near-white, or white) to be achieved in a particular application. Typical methods of surface preparation include abrasive blasting (wet or dry), acid etching, hand tools, power tools, high-pressure water blast, detergent wash, and solvent or detergent cleaning.

36.2.6 Application

Application methods vary from the traditional paint brush or roller to spray techniques, of which the latter is preferred for large-scale industrial applications. High pressure airless spray is commonly used. Exact procedures will vary with the paint system and application conditions, and should be based upon manufacturers' recommendations or those of a qualified consultant.

Brushing

Brushing is an effective, relatively simple method of paint application, particularly with primers, because of the ability to work the paint into pores and surface irregularities. Because brushing is slow, it is used primarily for smaller jobs, surfaces with complex configurations (edges, corners, cuts, and so on), or where overspray may constitute a serious problem. The other disadvantage of brushing is that it does not produce a very uniform film thickness.

Rolling

Rollers are best used on large, flat areas that do not require the surface smoothness or uniformity achievable by spraying. Rollers are also used in interior areas where overspray presents a cleaning and masking problem. However, the brush is preferred over a roller for applying primers because of the difficulties in penetrating pores, cracks, and other surface irregularities. In rolling, air mixes with the paint and leaves points for moisture to penetrate the cured film. Rolling is generally more suitable for topcoating a primer that has been applied by some other method.

Spray Painting

Spray painting of coatings results in a smoother, more uniform surface than that obtained by brushing or rolling, because these application methods tend to leave brush or stipple marks and irregular thicknesses. The most common methods of spray painting are conventional and airless.

The conventional spray method of spraying relies on air for paint atomization. Jets of compressed air introduced into the stream of paint at the nozzle break the stream into tiny droplets that are carried to the surface by the current of air.

Because large amounts of air are mixed with the paint during conventional spray application, paint losses from bounce back or overspray can be high. Such losses have been estimated to be as much as 30 to 40%.

However, some of the disadvantages of conventional air spray application are:

λ   It is slower than airless application;

λ   More overspray results than with other methods; and

λ   It is hard to coat corners, crevices, and so on, because of blowback.

Airless spray provides a rapid film build, greater flow into surface irregularities, and rapid coverage. In airless spraying, paint is forced through a very small nozzle opening at very high pressures to break the existing paint stream into tiny particles (similar to water exiting a garden hose). Because of the high fluid pressure of airless spray, paint can be applied more rapidly and at greater film builds than in conventional spraying. The high pressure paint stream generated by airless spray penetrates cavities and corners with little surface rebound. If airless is used, proper spraying technique is essential for quality paint films.

Some of the advantages of airless spray as compared with conventional spray include:

λ   Higher film builds are possible (heavier coatings);
     Less fog or rebound;
λ   Easier to use by the operator, because there is only one hose;
λ   Higher viscosity paints can be applied; and
λ   Easier clean up.

The disadvantages of the airless spray process include:

λ   Relies on dangerous high pressure;
λ   Fan pattern is not adjustable;
λ   Many more working parts that can cause difficulty;
λ   Higher initial cost than other forms of spraying; and
λ   Requires special care to avoid excessive build-up of paint that causes solvent entrapment, pinholes, runs, sags, and wrinkles.

36.2.7 Quality Assurance

To obtain the desired protection of a metal substrate from a coating system, it is important not only to choose the proper coating materials, but also to ensure that they are properly applied. On most jobs, such assurance is provided by a reputable paint force, or contractor, but on many other jobs, surveillance by plant personnel or thorough inspection by independent third-party organizations is done. In all cases, the rudiments of quality are the same:

λ Proper masking and protection of surfaces not to be blast cleaned or painted;

λ Removal of rust and contaminants, and suitable roughening of the surface;

λ Application of the specified coating system to the proper thickness;

λ Observation of application parameters, such as minimum and maximum temperatures, interval between coats, and induction times;

λ Verification of proper hardening or cure of the coating;

λ Testing to ensure that defects such as pinholes, skips, or holidays are minimized or avoided;

λ Tests for adhesion, color, gloss, and other parameters that may affect appearance or protective capability of the coating; and

λ Coating inspection requires training, experience, and familiarity with the instrumentation the inspector must use.

36.2.8 Justification

The economics of painting hinge on three factors: safety; metal replacement costs; and aesthetics. Because the corrosion rate is relatively low in many atmospheres, most current painting is done for aesthetic reasons. However, with the steady increase in materials costs and the depletion of our natural resources, the decision to achieve an adequate paint job to protect an investment is easily justified. Certainly, when safety of people is concerned (e.g., in maintenance of highway bridges), the expense is justified.

Justification for a paint system and its maintenance program can easily be determined by the methods described in Chapter 34. The cost of rigging, surface preparation, paint materials, inspection, and other factors must all be incorporated in the calculations. An evaluation of the procedures to be used must be made prior to implementation. A decision must be made as to whether to use company personnel (if available) or contractors. Individual contractors may operate on either a bid or a cost-plus basis. Many offer maintenance contracts as well.

It is more difficult to compute the effectiveness of a painting program. Only the cost per square foot per year is an adequate index. Expressing the cost as per cent of capital investment or of the maintenance costs is meaningless. Specifications for the entire job must be written. These will not in themselves assure a good job, however. A pre-job conference should be held to ensure a common understanding among all parties involved as to what is to be done and how.

Proper application of the paint system cannot be overemphasized. Trained application personnel and competent inspectors are essential. Expediency must not be allowed to compromise the job just to meet a schedule. The final job and its service performance will be no better than the attention given to application details.

Adequate inspection can greatly aid in obtaining a better job, but in the last analysis, the knowledge and integrity of the applicator crews are essential to a satisfactory paint or coating application.

Reference

1. ASTM Standard D16, "Definitions of Terms Relating to Paint, Varnish, Lacquer, and Related Products," *Annual Book of ASTM Standards* (Philadelphia, PA: ASTM).

Suggested Resource Information

C. G. Munger, *Corrosion Prevention by Protective Coatings* (Houston, TX: NACE International, 1985).

NACE Coatings and Linings Handbook (Houston, TX: NACE International, 1985).

NACE Surface Preparation Handbook (Houston, TX: NACE International, 1992).

37

Anti-Corrosion Barriers

This brief chapter discusses techniques which are a combination of two corrosion control techniques; a sort of a mix between changing materials of construction and interspersing a barrier layer between the material and the environment. The author has deliberately elected to consider anti-corrosion barriers as a separate entity from paints and coatings, for example.

This treatment of the subject derives from the fact that a change in material of construction can be partial rather than total. Corrosion may often be economically controlled by changing only the surface characteristics, e.g., by means of a fairly thick and substantial barrier (as distinct from paints, coatings, and thin electroplating or similar means). Indeed, in some cases, this anti-corrosion barrier may be applied only to a localized part of the equipment (e.g., the bottom of a tank or vessel).

Such a barrier may be organic (e.g., a lining of a plastic or elastomeric material), inorganic (e.g., glass, cement, metallic oxide), or metallic in nature. The following discussion is divided into metallic and non-metallic barriers.

37.1 Metallic Barriers

37.1.1 Plating and Metallizing

Because of the thin deposit formed, as well as a tendency to contain defects or holidays which would expose the non-resistant substrate, electroplating is not often used for corrosion control in distinctly corrosive services. More often, electroplating (or electroless plating, in the case of the nickel-nickel phosphide deposit) is used to minimize contamination of product or to improve lubrication, as in chromium-plating of rotating equipment, where the microcracked chrome plate tends to act as a reservoir for oil.

An important exception is tantalum plating, deposited from a molten salt bath. This has the unique characteristic of being pore-free. Tantalum-plated artifacts with a substrate of steel, copper alloys, etc. can be utilized in services which would rapidly attack the substrate. Typical applications include tantalum-plated thermowells and orifice plates.

Metallizing, as of zinc or aluminum, is effected by spraying molten metal on a steel substrate. The resultant surface is very porous but does effectively reduce the area of steel exposed, as in barge or tanker compartments handling iron-sensitive products.

37.1.2 Weld Overlays

A weld overlay is usually laid down in two or (preferably) three weld passes to deposit a surface of weld metal considerably more corrosion-resistant than the substrate. A familiar example is the deposition of type 308 (UNS S30800) or 316L (31603) stainless steel weld overlays on a carbon steel item to give it corrosion resistance equivalent to solid stainless steel.

The purpose of the multiple passes is three-fold. First of all, they override the dilution effect inherent in the first-pass melt mixture, which loses some of its alloy content by dilution with carbon steel. Secondly, they minimize the chance of a fault or holiday allowing access of the environment to the non-resistant substrate. Thirdly, they tend to assure that the final surface will have a chemical composition very nearly identical to that of the weld-rod itself. It should be noted that there are specialized modern techniques which can achieve these effects in a single pass weld overlay, but their use is restricted to shop fabrication under highly controlled conditions. For maintenance work in the field especially, the multi-pass weld overlay is to be preferred to single-pass work.

The weld overlay may be put down manually or under machine-controlled conditions, using any conventional welding process appropriate to the weld metal/substrate combination. There is a machine available which uses a submerged-arc technique to deposit an overlay while it rotates within a vertical cylindrical vessel.

Although weld overlays are intimately bonded to the substrate material, they do contain high residual stresses. The alloy selected for the overlay deposit should therefore be neither hot-short nor susceptible to environmental cracking in the environment of concern.

One of the weaknesses of a weld overlay is that it cannot be subjected to a quality assurance test for resistance to IGA (e.g., of a columbium [niobium]-stabilized or low-carbon deposit), as can the linings or claddings described below.

37.1.3 Linings

A metallic *lining* is fabricated by attaching a series of small segments of sheet or plate, usually by welding, to a less corrosion-resistant substrate metal or alloy. A substantial amount of linear welding footage is involved, depending upon the size of the segments relative to the total area of the vessel or part of the vessel to be lined.

Linings have an advantage over weld overlays to the extent that the sheet or plate can be subjected to quality assurance corrosion tests prior to welding in place. To a certain extent, they share the disadvantage inherent in a multiplicity of welds (i.e., the possibility of weld defects).

Another disadvantage can arise if there is a significant difference in thermal expansion between the alloy lining and the substrate. Such a difference can result in inordinately high stresses unless the lining is very carefully designed. Austenitic stainless steel linings in carbon steel vessels are notoriously subject to failure in service for this reason.

Vacuum service is very hard on lined vessels and, unless the vessel is specifically designed for the vacuum rating, can lead to premature failure, usually in the form of weld cracking. Once a leak occurs between the lining and the vessel wall, weld repair is difficult, if not impossible, because of foreign material trapped between them (a source of subsequent weld contamination).

Formerly, alloy linings have usually been more often specified for repair or refurbishing than for new construction. However, more recently a technique known as *wall-papering* has been used to attach very thin sheets of a corrosion resistant alloy in large sections, using plugwelds plus specially designed overlapping joints. This has proven successful particularly in flue-gas desulfurization units in the power industry. An even newer development combines the wall-papering and explosive cladding (see techniques on following page).

37.1.4 Cladding

Cladding is the preferred technique, from the standpoint of reliability, for the application of a corrosion-resistant barrier of a metallic nature on a less resistant substrate.

In this technique, a sheet or plate of a resistant alloy is *metallurgically bonded* to a less resistant substrate, whose function it is to provide mechanical strength. Usually, the cladding comprises 10 to 20% of the total wall thickness (e.g., one-eighth of an inch of alloy on a one-half inch of steel plate). The mechanical strength is predicated only on the thickness of the *substrate* material.

Unless there is a particular service requirement other than general corrosion resistance (e.g., having an *exterior* surface resistant to SCC), clad construction usually becomes economically competitive with solid alloy construction at approximately 16 mm (0.63 in.) to 19 mm (0.75 in.) total thickness.

The manufacturing processes for clad steel plate include rolling, brazing, and explosive bonding. The cladding material can be subjected to prior quality assurance tests. Nevertheless, the integrity of the metallurgical bond should subsequently be verified (e.g., by ultrasonic inspection). A partially disbonded cladding shares some of the same weaknesses as a lining (in vacuum service, for example).

Cladding is commonly employed to provide a corrosion-resistant barrier of copper or nickel alloys, stainless steels, or reactive metals on steel or low alloy steels. Because of the intimate metallurgical bond, differential thermal expansion is less of a problem than with linings.

There is, however, a problem in SCC service. Thermal stress relief is *not effective* for clad construction if there is a significant difference in thermal expansion between cladding and substrate (e.g., type 304 or 316 stainless cladding on carbon steel). One can go through the motions, e.g., two hours/inch of thickness at 900°C (1,650°F) and a slow cool. However, stresses will be reintroduced during the cooling process, due to the restraint imposed, and the cladding will still be susceptible to SCC.

To avoid or minimize this problem, high-performance stainless alloys (see Chapter 13) should be employed for cladding in services with a potential for SCC.

37.2 Nonmetallic Barriers

37.2.1 Organic

Organic barriers comprise both plastic or elastomeric sheet linings and heavy-duty coatings. The sheet linings most commonly employed are rubber linings of various kinds, and vinyl, polyolefin, or fluorinated plastics (e.g., FEP, PFA).

Whereas paints and coatings (see Chapter 36) are usually of a 3 to 15 mil (75 to 380 µm) thickness, heavy-duty films of 40 mil (1 mm) or more may be considered barrier materials. Among these, we find vinyl plastisols, fused fluorocarbons of 10 mil (0.25 mm or 250 µm) per coat, PFA coatings, flakeglass-filled polyesters, and vinyl esters. Currently, special furane coatings of up to 100 to 125 mil (2.5 to 3 mm) are also available.

37.2.2 Inorganic

Glass-linings of 60 to 80 mil (1.5 to 2 mm) have been previously mentioned. Cementitious barriers may be spray-applied on the interior of vessels. Brick-linings and carbon-brick are also employed to provide very thick anti-corrosion barriers on the interior of steel vessels. Although brick linings reduce the temperature of the process liquor should it gain access to the substrate, the conventional acid-brick is best employed in conjunction with a plastic or elastomeric membrane to protect the substrate when process liquors inevitably penetrate the brick lining.

38

Cathodic Protection

Past chapters have mentioned cathodic protection (CP). Although of primary interest to "pipe-liners" in the oil and gas industry, it behooves engineers in the chemical process industry to understand the principles and applications for interplant and intraplant underground piping, buried or mounded vessels, and water services.

38.1 Definition

Cathodic protection is the reduction or elimination of corrosion by making the metal structure wholly a cathode by means of an impressed current. The DC current employed may be generated by corrosion of a sacrificial anode or by the application of impressed current from an external source.

From the electrochemical standpoint, CP is effected by polarizing the local cathodes to a potential as negative as (or more negative than) the metallic anodes on the structure to be protected.

38.2 Principles

It has been previously noted that corrosion in natural environments is usually under cathodic control. In aqueous environments of high conductivity, this means that the principle controlling factor is the degree of polarization of the cathode.

In less conductive waters, and for underground applications in soil particularly, the effect of electrolyte resistivity is additive to the influence of driving potential between anodes and cathodes and their effective polarization.

The relationship between voltage (E), amperage (I), and resistance (R) is given by Ohm's Law:

$$E = IR \qquad (1)$$

where E is in volts, I in amperes, and R in ohms (or their values in milli-equivalents).

It must also be remembered that a series of individual resistances can be summed as:

$$R = r + r' + r''..., \qquad (2)$$

while *parallel* resistances (which apply particularly in soil) are summed as their reciprocals, where:

$$1/R = 1/r + 1/r' + 1/r''... \qquad (3)$$

Series resistance applies to structure/cable/anode connections, while parallel resistances apply to current paths in water or soil for the immersed anodes and structure, and to multiple anode connections supplying current to a pipeline.

38.2.1 Galvanic Anode Systems

When CP employs sacrificial anodes such as magnesium, zinc, or aluminum, the anodes function in a galvanic cell as a source of DC current. Corrosion is not stopped or slowed, but rather is transferred from the structure to the galvanic anodes, while a net current flow onto the structure effects CP. The electrochemical basis for CP by sacrificial anodes is illustrated in Figure 38.1.

Selection of galvanic anodes for a given system calls for consideration of three requirements:

(1) There must be sufficient driving potential between the anodes and the structure as cathode to overcome local anodes;

(2) The anodes must have enough contained energy to give an effective and economical life; and

(3) The anodes must have good efficiency, i.e., relatively little self- or auto-corrosion (which does not effectively contribute to the CP system).

Efficiency is usually expressed as a percentage but is practically evaluated in terms of amp-hours per pound (A-hr/lb). For example, zinc theoretically would provide 372 A-hr/lb (820 A-hr/kg)

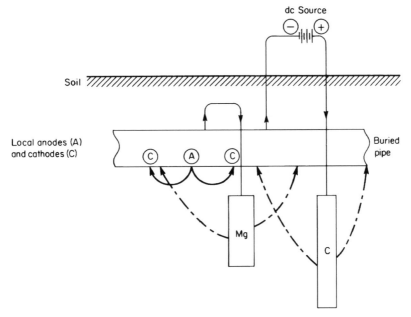

Figure 38.1 Galvanic (*left*) and impressed (*right*) current CP systems. Corrosion current: →; CP current: - - - - →.

but, at its 95% (0.95) efficiency, actually yields approximately 353 A-hr/lb (822 A-hr/kg). Magnesium is less efficient but cheaper than zinc and is preferred in most soil applications. It is subject to excessive autocorrosion in seawater (where zinc or aluminum are preferred). Replacement of galvanic anodes is predicated on figures such as these (i.e., how long an anode of given weight anode should survive).

38.2.2 Impressed Current Systems

Impressed current systems supply energy from an external source, and the system of structure and relatively inert anode(s) comprise a *sink*. The anodes are connected to the *positive* lead of the DC source. The situation is the reverse of a problem in electrolysis and is analogous to recharging the storage battery in an automobile.

Usually, the DC source is a rectifier connected to available AC power. However, other DC sources are also employed, such as storage batteries, motor-driven generators, windmills, fuel cells, thermoelectric devices, and solar panels.

Because of the way an impressed current system is connected, high-quality cable and carefully insulated cable connections are required. Otherwise, electrolysis *will* occur at weak points in the insulation. (Galvanic systems cathodically protect the substrate at such defects.)

38.2.3 Geometry Effects

In terms of where the current can most easily go, all CP systems are subject to the sort of limitations previously described under galvanic corrosion (Chapter 5). However, the practical aspects are somewhat more complex, especially in underground applications.

Figure 38.2 illustrates the current and voltage distribution related to a single anode for a section of pipeline. Galvanic systems often have n anodes per mile (or kilometer) of line, for example. The current distribution and attendant voltage changes are influenced by both distance of anode from the pipe and by soil resistivity. Figure 38.3 illustrates the effect of a low-resistivity soil such as clay on the same pattern.

A foreign metallic structure in the vicinity offers the same sort of low-resistivity path and also suffers electrolysis where the current *leaves* the foreign structure to go to the structure under deliberate protection (Figure 38.4). This could be corrected either with a separate anode system or, as shown, with a resistance bond (leaving the foreign structure unprotected).

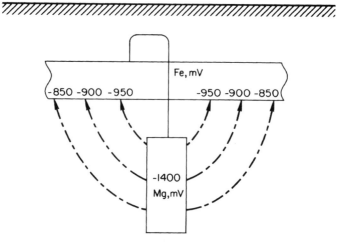

Figure 38.2 Current-potential distribution (galvanic anode).

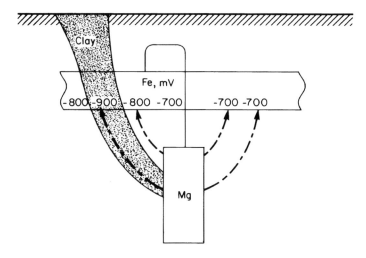

Figure 38.3 Effect of low-resistivity strata on current flow.

38.2.4 Criteria for Cathodic Protection

Any current from an external source which enters the structure is beneficial, as a rule. However, there are three commonly accepted criteria which indicate that adequate protection is being effected.

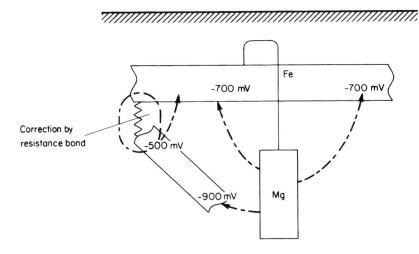

Figure 38.4 Effect of foreign structure.

Potental vs Environment

For steel in soil particularly (whose freely corroding potential usually lies between −300 to −700 mV to a copper:copper sulfate electrode), protection is thought to be achieved when the potential is lowered to minus −850 mV. If sulfate-reducing bacteria (SRB) are present, a further 100 mV reduction (to −950 mV) is recommended. Pipe-to-soil potentials (an integral part of design and operation) are read directly over the line, as close to its surface as possible (usually several feet, because of the earth cover). Potentials below approximately −1,100 mV tend to generate molecular hydrogen, which can waste protective currents as well as cause disbonding of coatings.

Other materials have different criteria; for example, −700 mV for lead (as in telephone cable sheath), and 1,000 to −1,200 mV for aluminum. For such amphoteric metals, *over-protection* is harmful because of attack by the alkali produced at the cathodic surface.

In other situations, different reference half-cells may be employed (e.g., silver: silver chloride in seawater). Then an algebraic correction must be made to the protective value. For example, the copper:copper sulfate electrode is +316 mV to the standard hydrogen electrode (SHE), while Ag:AgCl is +246 mV and the saturated calomel electrode (SCE) is +242 mV vs SHE. The protective value of −850 mV vs copper:copper sulfate is therefore −790 mV to silver:silver chloride and −534 mV vs SHE.

Potential Change

There are any number of reasons why the ideal protected potential may not be achievable in some particular installation. For this reason, a negative shift in potential of approximately 400 millivolts from the original is often considered indication of an acceptable level of CP.

Direction of Current Flow

Lastly, even with-less negative values observed than described above, low readings on a line are of little concern, provided it can be shown that the pipe is *gathering* current from all sides. Only a *discharge* of current from the structure to soil is cause for concern.

Two other criteria are sometimes employed:

(1)   Corrosion coupons (or other corrosion-detection devices) can be electrically attached to the structure, evaluation of which in the connected vs electrically isolated condition is a measure of the efficacy of corrosion control by the CP system; and

(2)   Vertical structures (e.g., well-casings) are designed on a current demand basis derived from a plot of voltage vs current, which is known as an E-Log I curve.

38.3 Design and Application

In designing a CP system, we must first know the current demand of the structure to be protected. (The soil resistivity is also a critical value for underground applications, determined by field measurement and calculation, using an applied current and voltage shift in the "four-pin" or Wenner method.)

In modern practice, buried or immersed structures are *coated* with a suitable high-resistance, high-dielectric system whose purpose is mostly to reduce the *area* of bare metal to be protected. For practical purposes, this means the metal exposed at original or potential holidays.

With underground piping, the system is electrically isolated at each end (to prevent egress of current) and a DC source is used to apply current until the pipe-to-soil potential reaches the desired criterion. From the current demand and soil resistivity, the necessary galvanic or impressed current system can be designed.

38.3.1 Local Conditions Affecting Design

Local conditions which will affect the design include:

(1)   The availability of power (for impressed current systems);

(2)   Location; for example, is this location suitable for installation, inspection, and maintenance? and

(3)   Presence of foreign structures in the vicinity and possible effects of other CP systems.

(4) In plant, particularly, such unusual conditions as past or potential future acid or chemical spills, coal-pile drainage, etc.

38.3.2 Factors Affecting Design

Some of the factors affecting design decisions include:

(1) Total current requirement: low current demand favors galvanic anodes while high current demand almost always requires impressed current systems.

(2) Variations in structure or environment: a well-coated pipeline in good soil will have very low demand, whereas bare steel in flowing water will have a high demand.

(3) The material to be protected (i.e., steel, lead, aluminum, stainless steel) must be considered in regard to protection criteria.

(4) The life requirement of the installation and degree of reliability and maintainability desired enter into the economics of the design.

(5) Suitable anode materials must be chosen (e.g., Mg, Zn, or Al for galvanic systems; silicon iron, carbon, platinized titanium, or other proprietary materials for impressed current systems).

Design details such as sizing of anode beds, etc. should be referred to a corrosion engineer with specialized training in the CP field.

38.4 Specialty Applications

Some specialty applications are well outside the interests of the process industry engineer (e.g., submerged oceanic lines; deep ground beds; marine applications for piling, hulls, ballast compartments and propellers; CP for rebar in concrete high bridges). However, in addition to buried lines and buried or mounded tanks or vessels, the chemical process industries engineer may become involved in water tanks, seawater flumes or piping, water-boxes for coolers and condensers, and processes themselves.

38.4.1 Crevice Corrosion and SCC

CP is sometimes employed against crevice corrosion and SCC (the latter usually via sacrificial metallic coatings). Modern practice is leaning towards potentiostatic control, derived from the laboratory apparatus/technique, to combat such phenomena, especially in the paper industry.

38.4.2 Anodic Protection

This technique, as previously noted, is the opposite of CP and can only be employed for special combinations of metal or alloy systems in specific appropriate environments. The most common applications lie in sulfuric acid storage in steel (to minimize iron contamination) and for type 316L coolers for hot, strong sulfuric acid.

Because of the ever-present danger of electrolysis in the event of loss of proper control, it is even more important than for CP that these systems be professionally designed and supervised.

Suggested Resource Information

J. H. Morgan, *Cathodic Protection*, 2nd Ed. (Houston, TX: NACE International, 1987).

NACE Standard RP0169, "Control of External Corrosion on Underground or Submerged Piping Systems" (Houston, TX: NACE International).

39

Inspection and Failure Analysis

It is necessary for reliable plant operation that process equipment be inspected on a regular basis in an effort to anticipate failures and prevent unscheduled shutdowns. But even with the best efforts, some failures will inevitably occur. This chapter addresses inspection schedules and techniques and the methodology of failure analysis. A section on information/data retrieval is also included.

39.1 Inspection

Some plant operations schedule an annual shutdown or *turnaround*, at which time it is convenient to inspect suspect equipment from the inside. But, whether or not such a shutdown schedule exists, it is helpful to divide equipment into categories which indicate both the urgency and timing for inspection. Inspection is usually divided into two categories, known as ETI (equipment test and inspection during shutdown) and OSI (on-stream inspection during operation).

39.1.1 ETI Planning

Equipment can be categorized as critical, routine, or long-term. Critical equipment is that in which any type of failure would cause a shutdown or pose a problem to operation or personnel safety. Routine equipment is that which requires only nominal attention for unexpected attack, while long-term equipment may be scheduled for inspection at prolonged intervals. With well-known operations, based upon either previous experience or with a thorough knowledge as to its low corrosivity, it is possible to schedule these categories as follows:

Category	Time to 1st Inspection, years
Critical	1
Routine	3 to 5
Long-term	10

If the process is not so well known or understood, one should schedule a first-year inspection in any event and then recategorize or reschedule, based upon initial findings. It should also be remembered that inspection may be required for reasons other than corrosion, such as debris or mechanical difficulties, which may affect the final scheduling.

Critical items are most often heat exchangers (because of the thin-walled tubes commonly employed) and reactors (because of possible localized exotherms, side-reactions or contaminants). Routine items are other vessels. Long-term items are exemplified by "day tanks" for feed streams or storage tanks for products.

39.1.2 On-Stream Inspection

OSI inspection does not require detailed scheduling, since it is performed while operations are in progress. It does, however, require careful planning as to what points in vessels or piping are to be monitored. OSI usually consists of corrosion monitoring by means of continuous read-out techniques and ultrasonic techniques (see further below) for wall thickness or cracking phenomena.

39.1.3 Inspection Techniques

The subject of recognition of corrosion phenomena was addressed in Chapter 6. It was emphasized that such phenomena can be divided into three groups: those which can be recognized by the naked eye; those which may require supplemental aids; and those which almost always demand sophisticated analysis. Actually, inspection utilizes basically the same techniques as does failure analysis up to the point where laboratory work (e.g., chemical analysis, metallography, etc.) is required. Among the techniques routinely employed are the following:

Visual Examination

Visual examination may be made by the naked eye. However, it is often fruitful to employ simple aids. A self-illuminated, low-power (10 to 30X) magnifying device is a useful tool. Dye penetrants, fluorescent or otherwise, may be used to help define and delineate defects, such as porosity or cracking. The Borescope[†] or similar tools permit visual inspection of the bore (internal surface) of small-diameter pipe or tubing, while glass-fiber optics permit one to see around corners or through small apertures. Developments in glass fiber optics, combined with video techniques, allow the interior of equipment to be observed remotely on a TV screen. A 100-degree field of view in a full-color, high-resolution video image can be achieved, useful for inspecting the inside of tanks, reactors, vats, large-diameter pipes, and other dark voids.

Magnetic Flux

For ferromagnetic materials, visual inspection may also be aided by use of an applied magnetic field and application of finely divided iron powder. The powder will orient itself in such a manner as to outline surface and/or subsurface defects.

Ultrasonics

High-frequency ultrasound will penetrate a metal, reflecting back from the far surface (to be visible on a video screen, if desired). Ultrasound may be used either for quantitative measurement of wall thickness or to detect cracks or other surface or subsurface defects. With high gain, even differences in grain size are detectable, so the instrument must be properly calibrated for detection of flaws of the anticipated nature and size.

Radiography

The two basic radiographic techniques are x-ray and gamma radiography, both of which are recorded photographically. In contrast to dye-penetrant techniques, radiography will detect subsurface defects such as porosity in welds. Radiography is not sufficiently sensitive to detect small cracks (e.g., SCC, HAC, LMC) unless they are of an orientation which will show a low-density picture (i.e., when they are parallel to the plane of radiography).

Modern developments in radiography, using a hand-held, battery-operated x-ray detector with a fluoroscopic display, to detect external corrosion of steel pipes up to 18 in. (450 mm) diameter without removing the insulation. A modification of this technique will measure internal corrosion of uninsulated pipe.

Eddy Current

Surface or subsurface defects will affect AC impedance, and advantage is taken of this characteristic in eddy current inspection of the bore of heat exchanger and other tubular parts. This technique is most applicable to non-magnetic materials (e.g., copper alloys, austenitic stainless steels, titanium). Ferromagnetic materials can be inspected, but require special techniques to compensate for the magnetism.

Acoustic Emission

While ultrasonic devices *apply* high-frequency sound for thickness measurements and flaw detection, acoustic emissions techniques listen for the high frequency sound emitted by crack-like defects under strain. With microsecond timing and computerized triangulation, acoustic emission equipment can indicate the location or general area of growing cracks or crack-like defects in a vessel. Typically, the test pressure is cycled during AE. This is a relatively expensive technique, but it is no longer a laboratory curiosity, being utilized industrially for detection of fatigue, as in structures, and environmental cracking, as in pressure vessels. AE is unique in that it tests the whole vessel or piping system and identifies areas in which more exact methods may be used to explore the problem further.

Corrosion Detection Devices

Either electrical resistance, electrochemical polarization, or hydrogen probes may be considered tools for OSI, indicating the relative rate of attack at particular times, as described in Chapter 9.

39.2 Failure Analysis

A failure is something which is unsuccessful, disappointing, or lacking to some degree. In the process and related industries, failure may consist in something as delicate (but not trivial) as a lapse in product quality (e.g., due to iron or other metal ion contamination). At the other extreme, there are fires, explosions, and/or toxic releases as a consequence of corrosion phenomena. There are all sorts of gradations in between, but we are most often concerned that we experience "leak-before-break" situations. The key concept is that failure is something *unexpected*, in contrast to the finite life one should always expect of a properly designed engineering structure.

Inevitably, there will be some failures, even with proper OSI and ETI programs. When they do occur, the inspection records are a good starting point for the investigation. In addition to the inspection techniques, the following methodology is normally followed.

39.2.1 Metallography

Metallography is augmented visual observation of a surface or of the internal crystal structure of a metal artifact and may be conducted at several levels of sophistication.

Low-Power Magnification

In addition to visual examination as previously described, a *shop-type microscope* is useful for low-power microscopy (e.g., from 5 to 60X). This instrument not only permits close examination of certain features of the failure, but is useful in recording, as by color photography, the surface features of corrosion and corrosion products, pitting, cracking, etc.

Metallurgical Microscope

The optical *metalloscope* or metallurgical microscope is a reflected light device (unlike the medical or biochemical microscope which uses transmitted light) designed for magnifications from approximately 50 to 3000X. A highly polished metal surface would simply act as a mirror, so the polished specimens must be etched, chemically or electrochemically, to diffuse the light and permit observation of metallurgical features. Most failure analysis work, unlike research-level metallography, is conducted at a few hundred magnifications to observe IGA, SCC, inclusions, and phases.

Scanning Electron Microscopy

While optical microscopy works from a surface polished and etched to define topographical features, the scanning electron microscope (SEM) indirectly observes *natural* surfaces of the material (or fracture surface) at magnifications which can be varied from 100 to 30,000X, for example. Using a video screen, the electron beams readily define the difference between ductile and brittle fractures, for example, features not readily determined by conventional optical microscopy. The SEM can also distinguish hot-short cracking from other forms of intergranular environmental cracking. Used with a special attachment such as a microprobe, qualitative and semi-quantitative analyses may be made at specific sites within the metal structure.

39.2.2 Corrosion Tests

Either quality assurance or service-related tests may be run in support of a failure analysis, in order to investigate the effects of composition, heat-treatment and/or process or service conditions. Both corrosion data and relevant supplementary and/or supporting information should be easily retrievable, as described in Chapter 39.4.

39.2.3 Chemical Analysis

Conventional wet analyses are employed on materials, corrosion products, and samples of process streams (e.g., for composition and contaminants). Modern, sophisticated techniques have excellent capability in analyzing metals and identifying corrosion products (e.g., x-ray diffraction). However, the analytical chemist needs guidance as to what to look for. Common mistakes include reporting "total acidity" when the titration includes iron salts or other corrosion products (or even hydrolyzable esters), confusing oxidizing cations with chromates in iodimetric analysis, and reporting total iron or iron corrosion products as magnetite without regard to origin or characteristics. Also important is to know when total acidity may be more important than pH, from the corrosion standpoint. A team consisting of a corrosion engineer and a chemist (or at least a close liaison effort) usually is necessary for gathering reliable data, let alone interpretation thereof.

39.3 Preventative and Predictive Maintenance

The ultimate purpose of both inspection and failure analysis in plant is to permit the maintenance department to develop proper scheduling of either preventive or preferably predictive maintenance.

Preventive maintenance is maintenance performed on a regular basis to forestall failure or to ensure attainment of the normal anticipated life. First applied to rotating equipment such as pumps and compressors, it has come to be applied to vessels and piping also.

Predictive maintenance is a more sophisticated extension of the preventive maintenance concept, quantifying corrosion rates and attempting to establish schedules for refurbishment or replacement of equipment, to minimize down-time and lost production.

Normally, the inspectors in the maintenance department or ETI group are perfectly capable of evaluating general corrosion (with loss of corrosion allowance) and the more obvious forms of localized corrosion. They should call on the corrosion engineer when there is any question about specific phenomena or when any type of cracking is detected.

The corrosion engineer's role consists in aiding such routine inspection as required, utilizing the equipment reports from ETI and OSI to predict the probable life of specific equipment, and running such laboratory and/or field corrosion tests as are required to monitor the life of equipment and the behavior of potential alternative materials of construction. Failure analysis, when necessary, is an essential part of this effort also.

Equipment records are the province of the inspectors. The *usefulness* of such records is limited by the extent to which knowledgeable people (like the corrosion engineer or other professionals) use them as an aid in deriving maintenance or replacement schedules.

For corrosive processes, such records are invaluable in the writing of regular process reviews intended to document the process conditions and the performance (with materials of construction and possible alternatives) of the equipment. A *simplified* flow diagram, kept up-to-date and including materials of construction and relevant specifications, is a valuable tool for maintenance planning.

39.4 Documents and Data/Information Retrieval

Inspection records are usually entered on a fairly concise form and filed both by unit and by type of equipment. Modern microprocessors permit the computerization of such records, with ready access to scheduled events (e.g., retubing exchangers, replacing seals, hydrotesting, inspection).

However, other documents (e.g., reports on materials selection, corrosion studies, failure analyses) are often of the essay type, as opposed to simple forms or the tabulation of relevant corrosion data. Two aspects of these documents are their format and the ability to retrieve the information.

Practical engineering documents, even *letters*, should follow an outline similar to that which follows (although letters, unlike formal memoranda, do not usually employ section headings):

> **Summary.** The summary (or the opening paragraph in a letter) should be a brief abstract of the reason for the study, with its findings and conclusions. This is probably as far as the "top brass" will read.

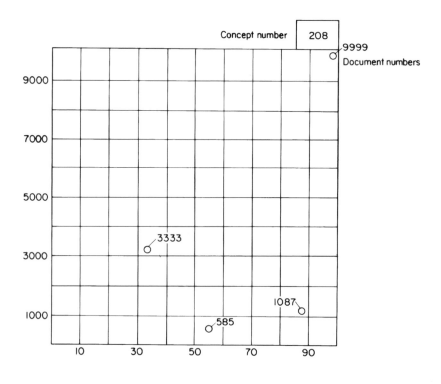

Figure 38.1 Concept card for information retrieval.

Introduction. This (or the second paragraph) is a more detailed statement of why the study was made, with background detail, and by whom it was made.

Discussion. The details of the history, methods, and observations during the study. (Technical details of the methodology should be relegated to an appendix, so as not to interrupt the flow of thought.) The text should be written so that the casual reader can scan the text by reading the *first sentence* of each paragraph only. (This follows naturally from a proper outline; paragraphs are intended to expand on the details of the lead sentence.) This is very helpful to middle management.

Conclusions. The findings of the study should be enumerated.

Recommendations. What the author suggests his client should do about whatever resolution of the problem has been made.

The author's peers will read his letters and reports in depth (including the detailed appendix), but this technique allows other levels of management and other disciplines to extract the necessary information either superficially or in depth, depending upon the extent of their interest.

Anybody can file letters and reports. The critical question is whether one can find, at some later date, what is in the files. In modern practice, conventional files are being replaced with retrieval systems.

Although some organizations are preoccupied with data retrieval systems just for quantitative material (e.g., corrosion rate data), a better approach is to handle all pertinent supporting documents as well. Data is not useful in itself; it is of value only when properly understood and applied, usually by the professional corrosion/materials engineer, to a specific problem. In fact, published data is often misapplied by those not trained in materials selection and corrosion control.

A more useful search can be effected the use of Boolean operators (i.e., AND, OR, NOT) employed in personal information management (PIM) systems. This approach with either a mainframe computer or PC allows one to search for "*this and/or that, not that*" or to search in specific time periods or exclude certain concepts (e.g., not to retrieve reports by a certain author; to exclude certain time periods).

The data-bases used in PIM's are text-oriented unstructured (free-form), quite different from conventional structured databases. There are a number of such proprietary systems available, of varying sophistication, cost and capabilities, which permit the systematic search and retrieval of either or both PC files or hard-copy (e.g., Persoft's IZE[†], Folio Corporation's VIEWS[†], MicroLogic's INFO SELECT[†]). Most such systems will hot-link directly to other programs used for word processing or spreadsheet programs. Some systems require the use of key-words, while more modern programs may allow full text or phrase searches.

When hard-copy is to be entered, the corrosion engineer must enter the desired information on what is essentially a PC version of the file card. This card, upon retrieval, then shows the relevant abstracted information and identifies the file location of the document(s).

A typical conceptual breakdown of subjects amenable to this approach in either computer programs or hard-copy include:

λ   Laboratory test data;

λ   Field test data;

λ   Physical and mechanical properties;

λ   Inspection data;

λ   Plant location or site;

λ   Type of equipment;

λ   Materials and coatings;

λ   Specific processes;

λ   Specific environments; and

λ   Specific phenomena.

[†] Trade name

The PC has opened new technology for managing both data and information. In addition, current computer technology allows hard-copy text to be extracted to computer media with up to 99% accuracy by using a scanner and Optical Character Recognition (OCR) package. (Present limitations include the inability to recognize hand-printed text or draft-quality print.) The scanned documents can easily be imported into the text-base for subsequent retrieval.

The effective corrosion engineer is one who has information readily accessible and is able to make decisions based on valid knowledge and prior experience—his or her own or others'. Such an information/data retrieval system is a valuable tool both for the experienced engineer and as a training tool for the engineer-in-training, providing rapid access to a wide variety of relevant material.

Suggested Resource Information

R. C. Anderson, *Inspection of Metals*, Vol. 1, *Visual Examination* (Materials Park, OH: ASM International, 1983).

Handbook of Case Histories in Failure Analysis, Vol. 1 (Materials Park, OH, ASM International, 1979).

Handbook of Case Histories in Failure Analysis, Vol. 2 (Materials Park, OH: ASM International, 1993).

D. J. Wulpi, *Understanding How Components Fail* (Materials Park, OH: ASM International, 1985).

Glossary of Acronyms

A

ABS	Acrylonitrile-butadiene-styrene
AE	Acoustic emission
AHF	Anhydrous hydrogen fluoride
AOD	Argon-oxygen-decarburization process
AP	Anodic protection

B

BD	Blowdown
BFW	Boiler feedwater
BFWMU	Boiler feedwater make-up

C

CAB	Cellulose acetate-butyrate
CBM	Constant boiling mixture
CI	Corrosion index
Cl⁻SCC	Chloride stress corrosion cracking
C.P.	Chemically pure
CP	Cathodic protection
CPVC	Chlorinated polyvinyl chloride
CR	Circulation rate
CTFE	Chlorotrifluoroethylene

D

DCI	Ductile cast iron
DM	Demineralized water
DOF	Definition of facilities
DO	Dissolved oxygen

D.O.T.	Department of Transportation
DOT	Definition of technology

E

ECTFE	Ethylene chlorotrifluroethylene
EMF	Electromotive force
ENP	Electroless nickel plating
EPDM	Ethylene-propylenediene methylene
ER	Electrical resistance
ER	Evaporation rate
ESC	Environmental stress cracking
ESCC	External stress corrosion cracking
ETI	Equipment test and inspection

F

FEP	Fluoroethylene propylene
FRP	Fiberglass-reinforced plastic

G

GMAW	Gas metal arc welding, MIG
GTAW	Gas tungsten arc welding, TIG

H

HAC	Hydrogen-assisted cracking
HAZ	Heat-affected zone
HBN	Hardness Brinell number
HE	Hydrogen embrittlement
HIBC	Hydrogen-induced blister cracking
HRC	Hardness Rockwell C

I

IGA	Intergranular attack

L

LMC	Liquid metal cracking (embrittlement)
LME	Liquid metal embrittlement (cracking)
LPRP	Linear polarization resistance probe
LSI	Langelier saturation index

K

KLA — Knife-line attack

M

MIG — (see GMAW)
MIP — Materials identification procedure
MU — Make-up water

N

NDT — Nil ductility temperature
NDTT — Nil ductility transition temperature
NHE — Normal hydrogen electrode

O

OSI — On-stream inspection

P

PA — Phenolphthalein alkalinity
PAIR — Polarization admittance instantaneous rate
PIM — Personal information management
POP — Pay-out period
PSI — Predictable saturation index
PWAT — Present worth after taxes
PVC — Polyvinyl chloride

R

RH — Relative humidity
ROI — Return on investment
RSI — Ryznar stability index
RTP — Reinforced thermoset plastic

S

SAW — Submerged arc welding
SCC — Stress corrosion cracking
SCE — Standard calomel electrode
SDI — Stiff-Davis index
SEM — Scanning electron microscope
SHE — Standard hydrogen electrode

SMAW	Shielded metal arc welding
SOHIC	Stress-oriented hydrogen induced cracking
SRB	Sulfate-reducing bacteria (bacterium)
SSC	Sulfide stress cracking
SSRE	Slow strain rate embrittlement
SWC	Step-wise cracking

T

TDS	Total dissolved solids
TH	Total hardness
TIG	(see GTAW)

U

| UDC | Underdeposit corrosion |
| UV | Ultraviolet |

V

| VOC | Volatile organic compounds |

End Notes

The purpose of this book has been to familiarize the reader with some fundamental concepts concerning materials of construction and related corrosion phenomena, as well as the relevant corrosion control techniques. This foundation in practical knowledge is intended to enable one to recognize a problem and the possible need for professional help. It should also arouse a further interest in corrosion science, technology, and engineering.

Technology is the reduction of scientific knowledge to engineering practice. While the specific fields of CP, painting, and water-treatment are continually developing specialists, there is a dearth of capable people developing as corrosion/materials engineers in the process industries. The industrial plant in chemical, petrochemical, and related processes requires guidance in a very broad field of materials, as regards design, operation, and maintenance.

Although many colleges and universities offer post-graduate degrees in corrosion and materials science, there is a notable paucity of educational institutions that have developed an undergraduate course in corrosion *control* engineering. A list of colleges and universities worldwide that include corrosion studies in their curriculum is published in NACE Internationals *Materials Performance* magazine as an "Annual Education Supplement."

The engineer who is seriously interested in specialization in corrosion control must, for the most part, be self-educated. Fortunately, there are a number of excellent education courses in corrosion control offered by NACE. They comprise Corrosion Fundamentals, Designing for Corrosion Control, Cathodic Protection, Protective Coatings and Linings, Microbiologically Influenced Corrosion, and Corrosion Control in the Oil and Gas Production.

These courses are given in an intensive (one-week) mode at several locations in the U.S. and Canada, several times a year. Some are available as home study courses, designed for people who want to attend the classroom version of a NACE course, but would find a home study program with a flexible schedule more convenient.

The Nickel Development Institute (NiDI) offers Materials Engineering Workshops. They give an overview of stainless steels and nickel-containing alloys used in the process industries with upgrading recommendations, high temperature corrosion, and fabrication and welding. These workshops, usually a half-day long, are held at locations throughout North America. Longer, more comphrehensive workshops are given overseas.

A very significant, new corrosion control resource is computer software. The CHEM•COR† program series is an interactive materials advisory (expert) system, a product of the NACE-NIST (National Institute of Standards and Technology) Corrosion Data Program with support of the Materials Technology Institute of the Chemical Process Industries (MTI) and the Nickel Development Institute. The series is available from NACE International.

CHEM•COR† prompts a question-and-answer dialogue between user and a knowledge/resource base similar to a consultation with a materials or corrosion specialist. Recommendations for selection of materials to use in shipping, handling, and storage of chemicals, mostly hazardous, are given based on service type, environment, and the presence of possible contaminants and contaminant concentration.

There are twelve modules in the CHEM•COR† series: concentrated sulfuric acid, acetic acid, formic acid, hydrogen chloride, chlorine, sodium hydroxide, phosphoric acid, once-through water,

---

† Trade name

hydrogen fluoride, dilute sulfuric acid, nitric acid, and ammonia. A step-by-step example of the use of CHEM•COR† is given in *Materials Performance* 32, 11 (November 1993), pages 14 and 15.

A more sophisticated version is CHEM•COR PLUS.† It allows the user to add information and customize the software program to meet individual or company specific needs.

Videotape programs are another important resource. *Corrosion Awareness* is a six-part series developed by Gulf Publishing Company and MTI. It also is available from NACE International. The videos are intended to train process plant people to recognize early signs of corrosion and be familiar with some materials, inspection, and joining methods commonly used in the CPI.

Titles in the *Corrosion Awareness* series are: Recognizing Corrosion, Materials Mix-Ups, Nondestructive Testing, Stainless Steels, Welding Quality, and Bolting.

• • •

It is my sincere hope that this volume will not only fill a practical information need but will encourage engineers and others in the chemical process industries to pursue further the important and interesting field of corrosion control.

Index

ABS (Acrylonitrile-Butadiene-Styrene), 91, 167, 393
Acetals, 169-170
Acetic acid, 2, 24, 78, 87, 143, 159, 161, 256, 258-262, 283, 291, 312, 343, 398
   handling and storage of, 161, 236, 259, 398
   metals vs, 258
   nonmetals vs, 260
Acid-gas scrubbing systems, 287-289, 339, 342
   inhibitors in, 340
   inorganic, 109, 163-166, 194, 239, 318, 320, 349, 351-352, 355-357, 365, 369
   oxidizing , 16, 56, 78, 94, 123-124, 129-130, 134, 140-143, 146, 150, 153, 156, 158-161, 167, 172, 174-175, 179-180, 191, 199, 204, 227-228, 230-231, 233, 235, 239-241, 248-255, 257, 259-260, 273, 279, 281, 286, 291, 293, 304-305, 307, 320, 340, 343, 347, 355, 386
   reducing, 67, 94, 123, 129, 141, 156-157, 159, 227-228, 233, 239, 250-253, 255, 263, 271, 304-305, 320
   (*See also* specific acids)
Acoustic emission (AE), 384, 393
Admiralty metal, 63, 151
Aggressiveness index (AI), 186
Alkaline environments, 177, 277-278
Alkaline salts, 161, 283, 292

Alloy designations for stainless steels, 131
Aluminum and its alloys, 9, 17, 20, 24, 31, 34, 57, 60, 62-64, 80, 86-87, 89, 95-96, 105, 107-110, 114, 119, 151-152, 158, 177, 205-206, 209, 213, 218, 220-222, 225, 228, 231, 233, 240, 243, 247, 251, 256, 258, 261, 263, 269, 271, 273, 277-278, 285, 287, 293, 298, 306, 311-312, 343, 348, 350, 357, 366, 372-373, 376, 378
   corrosion of, 2, 18, 55, 60, 65-67, 71, 108, 109, 132, 143, 159, 162, 178, 186-188, 205, 209, 213, 218, 223-227, 229, 239, 250-253, 259, 261, 264, 269, 292, 294, 309, 313, 319, 323, 342-343, 345-346, 371, 383
   numbering system for, 31, 121
   properties of, 23, 43, 107, 112, 120, 150, 165, 171, 190, 221, 295, 347
Amalgamation, 87, 311-312
Amines, 16, 109, 137, 141, 146, 150, 171, 197, 264, 277, 285, 287-288, 342-343
   metals vs, 287
   nonmetals vs, 288
Ammonia, 16, 20, 30, 89, 110, 140-141, 150, 152-153, 160, 202-203, 214, 219, 223, 228, 264, 277, 285-288, 305-306, 341, 343, 399

carbamates, 287
metals, 285
nonmetals, 286
stress corrosion cracking (SCC) of, 226
Ammonium hydroxide, 176, 285-287
Anhydrous hydrogen fluoride (AHF), 180, 242-246, 393
Annual costs, 334-335
Anodic protection (AP), 68, 236, 320, 379, 393
Anodic reactions, 339-340
Anodizing, 106, 108, 222, 281
ANSI (American National Standards Institute) piping code, 7, 30-31
Anticorrosion barriers, 320, 365, 369
Aqueous systems, application of, 340
Argon-oxygen-decarburization process (AO), 128-129, 135, 229, 393
ASME (American Society of Mechanical Engineers) BP&V code, 29
ASTM standards, 172, 222, 364
Atmospheric corrosion, 119, 125, 183, 217-219, 221, 224, 226, 264, 343
chemical factors in, 218
controlling factors in, 218
corrosion control in, 1, 220, 365, 398
environmental cracking, 33-34, 40-42, 46, 73-74, 77, 83-90, 94, 96, 98, 110, 141, 170, 243, 268, 278, 311-312, 320, 322-323, 367, 384, 386
factors in, 3, 23, 46, 210, 218
relative humidity, 218
special problems in, 224
types of atmospheres, 220
types of corrosion, 218
Austenite, 115-116, 124, 127, 129
Austenitic stainless steels, 10, 24, 34, 42, 51, 72-73, 77, 81, 83, 96, 102, 127-128, 133, 137, 229, 268, 279, 302, 307, 319, 384

Bacteria, 183, 191-192, 198-199, 202-203, 214, 376, 396
Barrier films, 317
Bases, 3
(See also Alkaline environments)
Bimetallic corrosion, 70
Black iron (carbon steel), 37, 80, 107, 115-116, 118, 121, 124, 159, 164, 171, 203, 218, 222, 236, 246, 264, 273, 278, 282, 285, 297, 306, 323, 343, 366-368
Blowdown (BD), 195-196, 199-202, 393
Boiler(s), defined, 6, 29, 43, 46, 153, 188, 191-192, 194-197, 393
Boiler water, 194
boiler feedwater (BFW), 188, 191-192, 393
boiler feedwater make-up (BFWMU), 188, 191-192, 194-197, 199-200, 393, 395
coagulation, 195
coordinated phosphate, 195
phosphate, 137, 195, 246, 292-293, 341
Brasses, 89, 151-152, 206, 223, 240, 271
Brazing, 34-35, 43, 88, 107, 152, 368
Brittle failures, 15
Bronzes, 34, 89, 147, 149, 151-153, 223, 240, 253, 270
aluminum, 9, 17, 20, 24, 31, 34, 57, 60, 62-64, 80, 86-87, 89, 95-96, 105, 107-110, 114, 119, 151-152, 158, 177, 205-206, 209, 213, 218, 220-222, 225, 228, 231, 233, 240, 243, 247, 251, 256, 258, 261, 263, 269, 271, 273, 277-278, 285, 287, 293, 298, 306, 311-312, 343, 348, 350, 357, 366, 372-373, 376, 378
phosphor, 147, 149-150, 152
silicon, 25, 32, 34, 67, 74, 86, 89, 108, 111-114, 119, 130-131, 139-140, 151-152, 174, 177, 195-196, 223, 229, 232, 234-235, 243, 247, 256, 258, 305-

306, 350, 378
tin, 57, 63, 105, 145, 147, 149, 151-
   152, 158, 232, 257, 274, 311
Buna S, N, 173
Butyl rubber, 174, 231, 260

CAB (cellulose acetate-butyrate), 167,
   393
Cadmium, 63, 86, 89, 221, 223, 295,
   298
   vs atmosphere, 223
Calcium hydroxide, 277, 283
Carbamates in ammonia, 287
Carbon, 27, 36-37, 67, 72, 77-80, 87,
   107, 111-112, 114-116, 118-
   121, 124, 127-129, 131, 134-
   135, 140, 142, 159, 164, 170-
   171, 173-176, 183, 192-193,
   197, 203-204, 218-219, 222,
   229-230, 235-236, 239, 246,
   249, 255-257, 259-260, 263-
   265, 268, 273, 275, 278, 281-
   282, 285-288, 297, 304, 306,
   323, 341, 343, 350, 366-368,
   378
Carbon dioxide, 36, 87, 140, 183, 192-
   193, 197, 219, 239, 263-265,
   268, 287, 304, 341
   metals, 263-264
   nonmetals, 265
Carburization, 303-304, 306, 308
Cast irons, alloy:
   8, 32, 60, 65, 67, 73-74, 94-
   95, 105, 111-114, 127, 139-
   140, 149, 187, 198, 203, 205,
   209, 212-213, 227, 229, 232-
   234, 236, 240, 243-244, 247,
   250-251, 256, 258, 264, 278,
   282-283, 285, 308
   chromium, 2, 20, 57, 77, 79, 89, 93,
   108, 113-114, 117, 120, 123-
   127, 129, 133-134, 139, 142,
   150, 206, 229, 247, 270, 278-
   280, 297-298, 305-307, 312
   copper, 5, 8, 16-17, 31, 34, 57, 59-
   63, 65-66, 72, 80, 84, 86-90,
   94-95, 100, 108-109, 114,

119-120, 123, 134, 139-140,
   145-146, 149-153, 158, 160,
   177, 197, 206-207, 210, 214-
   215, 220-221, 223, 225-226,
   228, 230-231, 235, 240-241,
   243, 245, 248, 253, 257, 259,
   263-265, 270-271, 274, 280-
   281, 286, 288, 291-294, 296,
   298, 311-312, 317, 326, 340,
   357, 366, 368, 376, 384
   molybdenum, 93, 113, 117, 120,
   123, 128-129, 133-135, 141-
   143, 158, 206, 241, 252, 259,
   297, 305, 307
   nickel, 2, 7, 10, 20, 31, 34, 37, 57,
   63, 79, 86-87, 89-90, 95, 98,
   112-113, 117, 120, 123, 125,
   127-129, 133-134, 137, 139-
   143, 145, 151, 203, 206, 226,
   230-231, 233-235, 241, 244,
   247-248, 250-251, 253, 257,
   259, 261, 265, 271, 274, 278,
   280-281, 283, 286, 288, 292-
   294, 298, 304-306, 308, 312,
   317, 368, 394, 398
   silicon, 2, 7, 10, 20, 31, 34, 37, 57,
   63, 79, 86-87, 89-90, 95, 98,
   112-113, 117, 120, 123, 125,
   127-129, 133-134, 137, 139-
   143, 145, 151, 203, 206, 226,
   230-231, 233-235, 241, 244,
   247-248, 250-251, 253, 257,
   259, 261, 265, 271, 274, 278,
   280-281, 283, 286, 288, 292-
   294, 298, 304-306, 308, 312,
   317, 368, 394, 398
   unalloyed, 111, 156-158
Castings, 4, 32, 34-35, 46, 107, 109,
   112, 131, 149-150, 229, 308
Cathodes, 57, 59, 64, 66, 72, 268, 271,
   371
Cathodic protection (CP), 1, 18, 65, 67-
   68, 72, 83, 85, 88, 90, 107,
   110, 147, 178, 210, 212-213,
   215, 221, 258, 295, 320, 351,
   371-379, 393, 397-398
   conditions affecting design, 377
   criteria for, 85, 375-376

definition of, 47, 49, 55, 128, 344,
393-394
design and application of, 377
galvanic anode systems, 372, 374
geometric effects, 374
impressed current systems, 90, 215,
371, 373-374, 377-378
local conditions affecting design,
15, 72, 214, 220, 371-372,
377
principles of, 115, 121
specialty applications of, 378
Cathodic reaction(s), 339-340
Caustic, definition of, 40, 66, 87-90,
123, 127, 133, 140-142, 146,
160, 177, 193, 195-196, 205,
277-283, 291-292, 319, 341
Caustic cracking, 133, 278-279
Caustic embrittlement, 87, 196, 278
Caustic gouging, 196
Caustic soda, 277, 283
Cavitation, 69, 72, 148, 153, 321
Cellulose acetate-butyrate (CAB), 167,
393
Cement, 178-180, 186, 207, 214, 219,
265, 283, 292, 365
portland, 178-180, 207, 292
specialty, 178-180, 207, 292
Cementation, 62, 109, 205, 312
Cementite, 116, 297
Ceramics, 176-177, 180, 232, 245, 254,
260, 265, 286
Chemical analysis, 37, 94, 382, 386
Chlorides, 89, 109, 127, 142, 184, 187,
190-191, 193, 200, 203, 207,
209-210, 213, 222-223, 225,
240, 246, 258, 263-264, 291-
292, 341
stress corrosion cracking (SCC) by,
18, 24, 68, 73, 83-84, 88-89,
91, 102, 129, 162, 226, 272,
326-327, 393-395
vs steel in water, 187
Chem•Cor, 1, 3, 6-7, 10-11, 15, 17, 30-
31, 37, 43, 45, 47-48, 50-51,
56, 64-66, 80, 84, 88-91, 93-
94, 96, 109-110, 112, 123-
124, 128, 130, 134, 136-137,

139-140, 143, 145, 157, 162-
169, 171-175, 177, 180, 183,
190, 193-194, 196-199, 201-
203, 205, 207-211, 213-214,
218, 220, 222, 228, 232, 237,
240, 251, 258, 261-262, 272,
275-276, 280, 283, 289, 312,
319, 327, 333, 339-342, 345-
359, 366, 371, 378, 382, 385-
386, 393, 397-399
Chemically pure (C.P.), 11, 51, 75, 91,
157, 162, 208, 228, 237, 254,
258, 260-261, 280, 289, 312,
327, 393
Chilled water, 193
Chloride stress corrosion cracking
(Cl⁻SCC), 88, 393
Chlorinated polyvinyl chloride
(CPVC), 168, 207, 232, 393
Chlorine, 9, 17-18, 156, 158-159, 161,
168, 191, 204, 219, 230, 273-
276, 293, 305, 398
specific materials in, 273-275
storage and handling of, 276
Chlorotrifluoroethylene (CTFE), 169,
393
Chromic acid, 146, 180, 231-232
Chromium-bearing, 142, 228, 230, 235,
241, 271, 286, 292
Chromium, 2, 20, 57, 77, 79, 89, 93,
108, 113-114, 117, 120, 123-
127, 129, 133-134, 139, 142,
150, 206, 229, 247, 270, 278-
280, 297-298, 305-307, 312
Circulation rate, 126, 130-131, 135-
137, 142-143, 150, 200-201,
393
Cladding, metallic, 24, 139, 146, 160,
367-368
Coatings, 3, 9, 18, 24, 40, 43, 88, 96,
110, 145, 147, 161, 163, 169,
171-172, 175, 178, 205, 212-
213, 215, 221-222, 225, 318,
320, 345-350, 352-357, 359,
361-362, 364-365, 369, 376,
379, 390, 398
Codes, 29-31
ANSI piping codes, 29

ASME pressure vessel codes, 29
Composition of materials in corrosion testing, 93
Compound interest, 330-331
Concentration cell, oxygen, 40, 59-60, 70
Concrete, 55, 178-179, 186, 197-198, 207, 214, 221, 265, 348-355, 378
  corrosion I through III, 178-179
  rebar corrosion, 178
Condensate, steam, 49, 65, 90, 192-194, 197, 204-206, 264, 343, 345
Constant Boiling Mixture (CBM), 239, 393
Construction and start-up, 1-3, 7, 17, 29, 32, 43, 45-50, 55, 75, 84, 86, 105, 107, 111, 115, 139-140, 147, 158, 163, 172, 175, 177, 179, 197-198, 202-203, 206, 209, 211, 220, 228, 231, 233, 236-237, 242, 246, 250, 254, 261, 281-282, 294, 311, 317, 320-321, 342, 344, 348-349, 365, 367-368, 387, 397
Consultants, 4
Contamination effects, 255, 358
Cooling-water systems, 197
  once-through, 191, 197-199, 201, 203, 208, 319, 340, 398
  recirculated, closed, 186, 196-199, 201-202, 319, 340
  recirculated, open, 186, 196-199, 201-202, 319, 340
Copper and its alloys, 5, 8, 16-17, 31, 34, 57, 59-63, 65-66, 72, 80, 84, 86-90, 94-95, 100, 108-109, 114, 119-120, 123, 134, 139-140, 145-146, 149-153, 158, 160, 177, 197, 206-207, 210, 214-215, 220-221, 223, 225-226, 228, 230-231, 235, 240-241, 243, 245, 248, 253, 257, 259, 263-265, 270-271, 274, 280-281, 286, 288, 291-294, 296, 298, 311-312, 317, 326, 340, 357, 366, 368, 376, 384

corrosion of, 2, 18, 55, 60, 65-67, 71, 109, 132, 143, 159, 162, 178, 186-188, 205, 209, 213, 218, 223-227, 229, 239, 250-253, 259, 261, 264, 269, 292, 294, 309, 313, 319, 323, 342-343, 345-346, 371, 383
types of, 5, 7, 27, 32, 38-40, 42, 45-47, 49, 69, 80, 84, 86, 88, 96, 111, 113-114, 124, 145, 147, 149-150, 165, 171, 180, 189, 194, 209-210, 215, 217, 220, 225, 246, 304-305, 321-322, 342, 347-348, 351-352
Corrosion, 1-7, 9-11, 15-18, 20, 23-25, 27-29, 31, 33-34, 36, 38-43, 45-51, 53, 55-63, 65-73, 75, 77, 79-80, 83-85, 87-89, 91, 93-102, 105, 107-110, 112-114, 118-121, 125-137, 139-143, 145-153, 155-159, 161-162, 172, 175-178, 180, 183, 185-189, 192, 197, 202, 204-205, 208-215, 217-230, 232-234, 237, 239-241, 243-244, 246, 248, 250-254, 257, 259, 261-265, 267-279, 283, 285-289, 291-296, 299, 301-303, 305-306, 309, 311-313, 315, 317-323, 325-327, 329, 332, 337, 339-346, 348-351, 356, 359, 363-368, 371-374, 377-379, 382-391, 393-399
by soil, 209, 374
by water and steam, 183
costs of, 2
  (see Costs of corrosion),
cracking phenomena, 18, 41-42, 69, 73, 322, 382
  (See also Environmental cracking)
dealloying corrosion , 42, 69, 72, 95-96, 101, 202, 240, 248
definition of, 47, 49, 55, 128, 344, 393-394
electrochemistry of metallic corrosion, 55
forms of, 42, 46, 69-71, 73, 75, 115,

171, 202, 269, 303, 362, 386-387
galvanic, 47, 59-60, 63-67, 69-72, 85, 87, 90, 98, 100, 102, 105, 107, 110, 129, 156-157, 159-160, 212-213, 217, 223, 235, 240, 253, 271, 281, 292, 318, 320, 342, 356, 372-374, 377-378
graphitic, 63, 72, 105, 111-112, 175-176, 205, 234-235, 240, 242, 246, 249, 255-257, 260, 265, 275, 281, 286, 288-289, 307, 342
high-temperature, 73, 114, 120, 125, 127-128, 153, 161, 292, 295, 301, 304-306, 309, 350, 398
intergranular (IGA), *****
localized, 18, 42, 46, 68-70, 72, 77, 79, 83, 95, 98, 134, 161, 202, 209, 213, 217, 228, 234, 268, 271, 294, 296-297, 318, 321, 325-326, 340, 342, 365, 382, 387
mechanisms, 3, 55, 165, 178
metallurgical phenomena, 45, 69
of steel by water, 187
oxygen concentration cell, 60
pitting, 40, 42, 46, 68, 70, 73, 78, 94-96, 98, 101, 109-110, 127, 129, 131, 134-136, 141, 202-203, 205-206, 210, 213, 217, 221-222, 240, 244, 249, 251, 267-269, 271, 273, 291-294, 320-321, 326, 340, 342, 356, 385
velocity effects, 47, 69, 72, 97, 234, 267
Corrosion allowance, 42, 321, 387
Corrosion characteristics in materials selection, 42
Corrosion control, 1-5, 11, 15, 17, 23, 43, 45-46, 49-51, 68-69, 212, 214-215, 220-221, 237, 262, 315, 317-318, 320, 322, 327, 329, 337, 365, 377, 389, 397-399

amenability of materials to, 43
basic considerations in, 15
by barrier films, 320
by change of environment, 318
by change of materials, 317
by electrochemical techniques, 100, 139, 165
design considerations in, 320
purpose, 1, 45, 93, 119, 136, 366, 386, 397
Corrosion coupons, 98, 101, 377
Corrosion-detection devices, 377
Corrosion fatigue, 18, 25, 29, 33, 40, 42, 73, 83-84, 130, 322
Corrosion index, 204, 393
Corrosion inhibition, 72, 202, 344
acid-gas scrubbing systems, 287-289, 339, 342
aqueous systems, 340
electrochemistry of, 55, 339
nonaqueous environments, 342-343
environments, 3, 7, 17, 34, 41-42, 46-47, 50, 352, 354, 356, 371, 379, 390
refrigeration brines, 341
Corrosion inhibitor(s), definition of, 286, 339, 343-344
Corrosion mechanisms, 3, 55
Corrosion phenomena, 6, 28, 40-42, 56, 69, 96, 217, 306, 382, 384, 397
Corrosion probes, 99-100
electrical resistance, 100, 384, 394
electrochemical (PAIR), 100
hydrogen, 17-19, 42, 56, 58, 61-63, 66, 73, 83, 85, 90, 95, 99-100, 109, 118, 120, 123-124, 126-127, 140-141, 150, 155-156, 159-160, 179-180, 187, 191, 193, 196, 202, 204, 219, 222-223, 227-228, 235, 239, 241-243, 249-251, 253, 256, 262-264, 267-272, 277, 287-288, 292, 295-299, 304-305, 307, 343, 376, 384, 393-396, 398-399
Corrosion racks, 98, 100
Corrosion rate calculations, 101

Corrosion testing, 46, 93, 102, 344
coupon evaluation, 101
coupons, 94-101, 360, 377
field tests, 98
laboratory, 9, 46, 48, 59, 93, 96-98, 101-102, 114, 160, 176, 190, 240, 242, 245, 247, 252, 254, 270, 280, 320, 379, 382, 384, 387, 390
materials characteristics in, 95
materials factors in, 93
methodology, 97
phenomena-related, 96
quality assurance, 50, 80, 86, 93, 96, 326, 362, 367-368, 386
racks, 98-100
service-related, 97, 386
Corrosive service, 36, 45, 106, 145
Costs of corrosion, 1, 2
cost as a factor in materials selection, 23
hidden, 2
irreducible, 1
reducible, 2
(*See also* Economic comparisons)
Cracking phenomena, 18, 41-42, 69, 73, 322, 382
(*See also* Environmental cracking)
Creep, 120, 146, 164, 302-303
Crevice corrosion, 25, 46, 70, 94, 98, 101, 129, 134-136, 156-158, 162, 259, 275, 294, 320, 323, 340, 379
Crevices in design, 321
Cryogenic conditions associated with corrosion, 225
Cupronickels, 139, 153, 206, 271, 280
Curie point, 139-140

Data retrieval, 381, 389, 391
Dealloying, 42, 69, 72, 95-96, 101, 202, 240, 248
Decarburization, 297, 303-304
Definition of facilities (DOF), 47, 49, 393
Definition of technology (DOT), 47-48, 394

DOT (Department of Transportation), 31, 394
Demineralized water (DM), 191-192, 252, 393
Depreciation, 2, 332-335
Design considerations in corrosion control, 320
Dezincification, 72-73, 96, 151, 206, 240, 264
Discounted cash flow (DCF), 333-334
Discounting, 331-332
Dissolved oxygen (DO), 2-3, 19, 23, 45-46, 56, 58, 66, 78, 98-100, 109, 124, 150, 152, 160, 163-165, 168, 175, 184, 186-188, 194, 197, 204-206, 211, 228, 239-240, 248, 252-253, 257-258, 261, 264, 267-269, 292, 295, 301, 304, 319, 324-325, 339, 348, 354, 361, 366, 385, 388-389, 393
Distilled water, 72, 190-192, 252, 263, 312
Documents and information retrieval, 387-388
Ductile cast iron, 111-113, 233-234, 236, 393
Duplex stainless steels, 121, 130

Economic comparisons, 329, 336
Eddy current, 384
Elastomers, 49, 95, 173-175, 207, 236, 239, 249, 255-257, 260, 265, 275, 286
Electrochemical techniques, 3, 43, 97, 100, 139, 215, 221, 320
Electrochemistry, 55, 339
of corrosion inhibition, 339-340
of metallic corrosion, 55
Electrical resistance, 100, 384, 394
Electroless nickel plating (ENP), 139-140, 394
Electrolysis, 36, 67-68, 209, 373-374, 379
Electromotive force (EMF), 64, 66, 156, 394
Electromotive resistance (ER), 100,

200-201, 394
Electromotive series, 61-62, 150, 239,
304
Elongation, 24, 26, 108, 112
Embrittlement, 56, 73, 84-85, 87, 95-
96, 120, 127, 129, 141, 156,
160, 196, 244, 249, 268-269,
278, 295, 302, 307, 394, 396
blue (temper), 302
caustic, 40, 66, 87-90, 123, 127,
133, 140-142, 146, 160, 177,
193, 195-196, 205, 277-283,
291-292, 319, 341
hydrogen, 17-19, 42, 56, 58, 61-63,
66, 73, 83, 85, 90, 95, 99-100,
109, 118, 120, 123-124, 126-
127, 140-141, 150, 155-156,
159-160, 179-180, 187, 191,
193, 196, 202, 204, 219, 222-
223, 227-228, 235, 239, 241-
243, 249-251, 253, 256, 262-
264, 267-272, 277, 287-288,
292, 295-299, 304-305, 307,
343, 376, 384, 393-396, 398-
399
slow strain rate embrittlement
(SSRE), 295-296
475 °C, 129
Endurance limit, 28-29, 73, 145
Energy considerations, 15, 20, 46
Environmental aspects, 19
Environmental cracking, 33-34, 40-42,
46, 73-74, 77, 83-90, 94, 96,
98, 110, 141, 170, 243, 268,
278, 311-312, 320, 322-323,
367, 384, 386
by ammonia, 89, 141, 150, 153
by caustic, 160, 195, 279-281
by chlorides, 109
by hydrogen sulfide, 73
by mercury, 90, 141, 151-152, 312
by polythionates, 270
of aluminum, 20, 24, 34, 96, 105,
108-109, 209, 218, 258, 312,
343
of copper alloys, 90, 149, 214, 226
of lead alloys, 145
of magnesium alloys, 86, 105-106,

232, 240, 251
of metallic materials *****
of nickel alloys, 139
of plastics, 49, 84, 163-165, 168,
170, 239, 255
of stainless steels, 132, 142, 206,
264, 398
of steel, 18, 24, 62, 65-66, 68, 86,
107, 112, 114-115, 117-118,
121, 155, 159, 187, 217-218,
224-225, 251-252, 296, 302,
307, 319, 342-343, 345, 351,
360, 366, 368-369, 383
of titanium, 64, 78, 156, 162, 265,
271
of zirconium, 15
design against, 323
EPDM (ethylene-propylene diene
monomer), 174, 207, 260,
293, 394
Epoxy resins, 281-282
EPR (ethylene-propylene rubber), 174
Equipment test and inspection (ETI),
381, 385, 387, 394
ETI planning during shutdown, 381
inspection techniques, 1, 3, 16-17,
19, 25, 29, 50, 74, 101, 298,
325-326, 362-364, 368, 377,
381-388, 390-391, 394-395,
399
on-stream inspection (OSI), 50,
381-382, 395
Equivalent uniform annual cost, 23,
317, 330, 334
Erosion-corrosion, 18, 27, 69, 72, 247,
321, 340
Evaporation rate (ER), 100, 200-201,
394
Exfoliation, 80, 95-96, 109-110
Explosion hazards, 15
External stress corrosion cracking
(ESCC), 88, 225, 394

Fabrication characteristics, 32
Failure analysis, 3, 69, 381-382, 384-
387, 391
Faraday's law, 64-65

Fatigue, 18, 25, 28-29, 33, 40-42, 73, 83-84, 130, 146, 322, 384

Fatigue strength, 28

Ferrite, 37, 94, 115-116, 124, 129

Ferritic stainless steels, 27, 74, 80, 124, 126-127, 278

Flat products, 33

Fluorocarbon plastics, 168-169

Fluorinated rubber, 245, 255, 260, 275, 369

Fluoroethylene propylene, 394

Fiberglass-reinforced plastic (FRP), 164, 172, 180, 197-198, 203, 207, 214, 233, 236, 242, 249-250, 255-256, 282, 394

Field corrosion testing, 46, 98

Fire-control water, 49, 193

Fire hazards, 15

Forgings, 32

Formic acid, 78, 256-258, 262, 398

Forming characteristics, 34

Fresh water, 189-191, 197-198, 205, 207, 339

Fretting, 47, 69, 72, 135

FRP (fiberglass-reinforced plastic), 164, 172, 180, 197-198, 203, 207, 214, 233, 236, 242, 249-250, 255-256, 282, 394

Fuel ash corrosion, 305

Furnaces, 308

Fusion zones (FZ), 39

Galvanic anode systems, 372

Galvanic corrosion, 47, 59-60, 65-67, 69-70, 72, 102, 217, 253, 292, 374

Galvanic series, 63-64, 66, 72, 105, 107, 156

Gamma loop, 124-126

General (uniform) corrosion, 69-70, 73, 87, 94-95, 127, 131, 145, 217, 244, 267, 293-294, 311, 321-322, 340, 346, 368, 387

Glass, 72, 147, 159-160, 170-171, 176-177, 187-188, 232, 235, 241, 245, 249, 254, 257, 260, 265, 275, 281-282, 293, 322, 365, 383

Glass-lined equipment, 176, 281, 286

GMAW (gas metal arc welding, MIG), 36-37, 108, 156, 394-395

Gold, 56-57, 63, 123, 161, 230, 235, 241, 245, 249, 254, 271, 281, 286, 304-305

Graphite, 63, 72, 105, 111-112, 175-176, 205, 234-235, 240, 242, 246, 249, 255-257, 260, 265, 275, 281, 286, 288-289, 307, 342

Graphitic corrosion, 73, 95, 114, 205, 213, 264

Graphitization, 73, 118, 120, 302, 307

GRP (glass-reinforced plastic), 172

GTAW (gas tungsten arc welding, TIG), 37, 108, 156, 394, 396

Half-cells (standard), 376

Halogens, 142, 161, 175, 228, 276, 305

Hardness, 18, 27-28, 33, 85, 107, 115-120, 124, 139-140, 145, 152, 164-165, 168, 184, 186, 189-190, 192, 195, 199, 201, 205, 270, 298, 348, 353, 394, 396

Hardness, of water, 190

Hardness Brinell number (HBN), 28, 394

Hardness Rockwell C (HRC), 28, 85, 88, 270, 298, 394

Hazardous materials, 15, 17, 31
  specific hazards and related protective measures, 17-18
  (See also Explosion hazards; fire hazards)

Heat-affected zone (HAZ), 39, 72, 77, 79-80, 94-95, 108, 110, 141, 159, 241, 307, 394

Heat exchanger(s), 24, 32, 101, 149, 158, 171, 175, 318, 321-323, 384
  seal welding, 323
  strength welding, 324
  tubewall temperature, 88, 206, 326
  venting of top tubesheet, 326

Heat exchanger tubing: 18, 24-25, 32, 93-95, 99, 101, 140-141, 149,

158, 163-164, 167-171, 175, 312, 318, 321-323, 325
quality control, 325
safe ending of, 327
Heat-resistant alloys, 306
Heat treatment, 25, 33, 42, 74, 77, 94, 112, 114-115, 118-121, 124, 126-127, 140-141, 149, 222, 229, 270
High performance stainless steels, (see Stainless steels, specialty)
High-temperature phenomena, 301
  alloying elements, effects of, 37, 80, 113-114, 117-118, 120, 128, 133, 297, 306
  effects of, 42, 48, 98, 118, 157, 183, 197, 211, 260, 269, 306, 311, 377, 386
  corrosion high-temperature phenomena, 69, 73, 301-303
  fuel ash, 305
  halogens, 142, 161, 175, 228, 276, 305
  internal stability, 302
  metal behavior, 301
  molten metals, 159, 306
  molten salts, 123, 301, 305
  oxidation reduction, 100, 304
  sulfidation, 305
  surface stability, 303
Homogeneity, effect of, in corrosion testing, 33, 94, 96
Hot-short cracking, 25, 37, 42, 74, 152, 386
Hydriding, 156, 228, 244, 271, 281
Hydrochloric acid, 17, 80, 109, 141, 146, 159, 161, 227, 239-241, 291, 342
  inhibited, 17, 153, 157, 199, 203, 206-207, 240, 251, 285, 340, 342
  metals vs, 243-245
  nonmetals vs, 245-246
  storage of, 68, 140, 161, 218, 234, 236, 258-261, 287-288, 398
Hydrogen-assisted cracking (HAC), 18, 73, 83, 85-90, 126, 141, 157, 243-244, 295-296, 383, 394

Hydrogen blistering, 243, 270, 295-296
Hydrogen embrittlement (HE), 56, 249, 269, 394
Hydrogen phenomena, 18, 295
  in high temperature, 292
  practical guidelines for, 298
Hydrogen-induced blister cracking (HIBC), 295, 394
Hydrogen sulfide, 17, 19, 73, 85, 160, 179, 193, 202, 219, 223, 227, 239, 250-251, 264, 267-272, 287-288, 296-297, 305, 307, 343
  general corrosion in, 267-268
  hydrogen-assisted cracking in (SSC), 18, 73, 83, 85, 295-296, 394
  hydrogen effects in, 268
  localized corrosion in, 83, 268
  pyrophoric products in, 272
  specific materials, 4, 17, 42, 75, 86, 165-166, 173, 202, 211, 221, 240, 243, 247, 250-251, 256, 258, 263, 269, 273, 278, 285, 287

Impact strength, 119, 127, 156, 168
Impact tests, 74
Incoloy 800 (N08800), 133, 318
Incoloy 825 (N08825), 79, 89, 134-136, 248, 253, 259, 318
Inconel 600 (N06600), 37, 80, 142, 230, 241, 244, 248, 253, 260, 271, 274, 280, 286
Inconel 625 (N06625), 80, 142-143, 207, 224, 230, 248, 254, 260, 271
Indexes*****
  (see Scaling indexes in corrosion by water),
Information retrieval, 387-388
Information sources, 207
Inhibition, corrosion, 1, 10, 72, 100, 187-188, 202, 318, 341, 344
  (see Corrosion inhibition)
Inhibitors, corrosion, 202, 293, 320, 337, 339-340, 342-343, 346

(*See also* Corrosion inhibitors)
Inoculated cast irons, 112
Inorganic acids, 239, 352, 355
Inspection and failure analysis, 1, 3, 16-
    17, 19, 25, 29, 50, 74, 101,
    298, 325-326, 362-364, 368,
    377, 381-388, 390-391, 394-
    395, 399
Insulation, corrosion under: 18, 20, 49,
    89, 224-226, 305, 374, 383
    of stainless steel, 73, 123-124, 270,
    286
    of steel, 18, 24, 62, 65-66, 68, 86,
    107, 112, 114-115, 117-118,
    121, 155, 159, 187, 217-218,
    224-225, 251-252, 296, 302,
    307, 319, 342-343, 345, 351,
    360, 366, 368-369, 383
Interest, compound, 7, 29, 77, 80, 93,
    125, 149, 152, 155, 160, 163,
    175, 186, 220, 243, 247, 253,
    256, 260, 265, 319, 330-331,
    334, 371, 389, 397
Intergranular attack (IGA), 51, 69, 72,
    77, 80-81, 94-97, 102, 109-
    110, 126-128, 134, 141-142,
    146, 202, 217, 222, 229, 293,
    367, 385, 394
Intergranular corrosion, 42, 50, 77, 79-
    80, 259, 288
    of aluminum, 20, 24, 34, 96, 105,
    108-109, 209, 218, 258, 312,
    343
    of brass, 96, 151
    of bronze *****
    of nickel alloys , 139
    of stainless steels, 132, 142, 206,
    264, 398
    of steel, 18, 24, 62, 65-66, 68, 86,
    107, 112, 114-115, 117-118,
    121, 155, 159, 187, 217-218,
    224-225, 251-252, 296, 302,
    307, 319, 342-343, 345, 351,
    360, 366, 368-369, 383
    of zirconium, 15
Internal stability, 302
Investment, return on (ROI), 329, 332-
    333, 363-364, 395

Iron and steel, 5, 8, 111, 121, 187, 218,
    222, 229, 232-233, 240, 243,
    247, 251, 256, 258, 263, 269,
    273, 278, 285, 287, 293, 312
    (*See also* Steels)
Iron-carbon phase diagram, 115, 124
Irons, case (see Cast irons)

Knife-line attack (KLA), 79, 395

Laboratory corrosion tests, 96
Langelier saturation index (LSI), 184-
    186, 189, 394
Lead and its alloys, 36, 42, 58-59, 63,
    84, 86, 89, 109, 120, 145-147,
    151, 159, 177, 186, 208, 213,
    223, 228, 230, 232, 234, 240,
    243, 248, 253, 257, 259, 263-
    264, 268, 271, 274, 281, 283,
    288, 293, 306, 311, 340, 357,
    360, 367, 373, 376, 378, 389
    corrosion resistance of, 10, 108,
    149, 159, 218, 271, 278, 287
    forms of, 42, 46, 69-71, 73, 75, 115,
    171, 202, 269, 303, 362, 386-
    387
    types of, 5, 7, 27, 32, 38-40, 42, 45-
    47, 49, 69, 80, 84, 86, 88, 96,
    111, 113-114, 124, 145, 147,
    149-150, 165, 171, 180, 189,
    194, 209-210, 215, 217, 220,
    225, 246, 304-305, 321-322,
    342, 347-348, 351-352
Light metals, 105, 233, 243, 247, 251,
    258, 263, 273, 278, 285, 287
Linear polarization resistance probe
    (LPRP), 394
Linings, 3, 168-169, 174-175, 318, 355,
    364, 367-369, 398
    metallic, 24, 55-56, 59, 72-73, 77,
    80, 83, 86, 88, 91, 105, 145,
    150, 160, 176-177, 203, 213,
    221-222, 261, 272, 304-305,
    318, 320-321, 365, 367-368,
    371, 374, 379
    nonmetallic, 3, 72, 93-94, 96, 163,

175, 230, 232, 235, 241, 245, 254, 257, 260, 265, 365, 369

Liquid metal cracking embrittlement (LMCE), 394

Liquid metal cracking (LMC), 18, 73, 84-90, 120, 141-142, 151-152, 271, 311-312, 383, 394

Liquid metal embrittlement, 73, 84-85, 394

Literature sources, 9

Localized corrosion, 18, 42, 69-70, 83, 95, 98, 134, 161, 202, 268, 294, 325-326, 340, 387

Local conditions affecting design, 377

Machining characteristics in materials selection, 41

Magnesium, 57, 63, 66-67, 86-88, 105-109, 178-179, 184, 189, 191-192, 195, 209, 221, 232, 240, 243, 247, 251, 263, 273, 278, 283, 292-293, 372-373

Magnesium anodes, 67

Magnetic particles, 383

Magnetic properties, 25

Magnetite, 59, 100, 115, 386

Maintenance, 1-4, 10, 19, 23, 45-46, 101, 208, 272, 327, 335, 348, 353, 355, 363-364, 366, 377, 386-387, 397 *****

predictive, 97, 386-387

preventive, 17, 19, 386-387

Make-up water (MU), 192, 194, 197, 200-201, 395

Malleable iron, 112

Martensite, 116-117, 124, 126

Martensitic stainless steels, 125

Materials of construction for corrosive service, optimum, 1-3, 282, 311, 317, 320-321, 365, 387

Materials characteristics in corrosion testing, 50, 95

Materials conservation, 15, 20

Materials considerations, 45

Materials-environment interactions, 46

Materials factors in corrosion testing, 93

Materials identification procedure (MIP), 50, 395

Materials selection, 1, 3-5, 7, 11, 17-18, 20, 23, 42-43, 45-51, 74, 208, 214-215, 220, 231, 237, 261, 289, 388-389

considerations in, 15, 93, 320, 343

and environment interactions, 83, 303

factors in, 3, 23, 46, 210

interactions, 45-46

procedures and communications in, 45, 47, 50, 166, 336

for specific equipment, 46-47

Mechanical properties in materials selection, 4, 23, 25, 27, 31, 55, 107, 112, 115-116, 118, 120, 128, 139, 145, 150, 158, 163-166, 169-171, 302, 390

Melting point or range, 25, 34, 74, 115, 155, 158-159, 277, 301, 304

Mercury, 16, 73, 84, 86-87, 90, 109, 141, 151-152, 159, 294, 306, 311-313

amalgamation, 311-312

effects of, 42, 48, 98, 118, 157, 183, 197, 211, 260, 269, 306, 311, 377, 386

environmental cracking by, 312

liquid metal cracking (LMC), 311-312

Metallic barriers, 365

Metallic materials, 83

Metallizing, 365-366

Metallography, 382, 385

Metallurgical phenomena, 45, 69, 74

Methanation by hydrogen, 120, 296-298

Methodology:, 2, 96-97, 381, 385, 389

in corrosion control, 15, 51, 320, 397-398

in corrosion testing, 97-98

MIC (Microbiologically Influenced Corrosion), 202

Microscopy, 70, 385-386 (*See also* Metallography)

MIG, 36, 394-395 (*See also* GMAW)

Modulus of elasticity, 25-26

Molten metals, 159, 306

Molten salts, 123, 301, 305
Molybdenum, 93, 113, 117, 120, 123, 128-129, 133-135, 141-143, 158, 206, 241, 252, 259, 297, 305, 307
Monel (Alloy 400), 63, 80, 87, 90, 140-141, 153, 198, 206, 233, 244, 246, 248, 250, 253, 259, 265, 276, 280, 312

NACE International (formerly National Association of Corrosion Engineers), 5, 9-11, 51, 71, 75, 91, 102, 110, 121, 132, 137, 143, 153, 162, 180, 208, 237, 261-262, 272, 276, 283, 289, 299, 309, 327, 337, 344-345, 360, 364, 379, 398-399
NACE standard, 85, 91, 97, 102, 270, 272, 336-337, 379
Nelson curves for hydrogen vs steels, 297
Neoprene rubber, 245
Neutral salts, 178, 291
Nickel (Alloy 200) and its alloys, 2, 7, 10, 20, 31, 34, 37, 57, 63, 79, 86-87, 89-90, 95, 98, 112-113, 117, 120, 123, 125, 127-129, 133-134, 137, 139-143, 145, 151, 203, 206, 226, 230-231, 233-235, 241, 244, 247-248, 250-251, 253, 257, 259, 261, 265, 271, 274, 278, 280-281, 283, 286, 288, 292-294, 298, 304-306, 308, 312, 317, 368, 394, 398
  chromium-bearing, 142, 228, 230, 235, 241, 271, 286, 292
  chromium-free, 271
  nickel-chromium-molybdenum alloys, 121, 142
  molybdenum alloys, 206
  nickel-molybdenum alloys, 34, 74, 141
Nil ductility temperature (NDT), 18, 395
Nil ductility transition temperature (NDTT), 27-28, 74, 112, 114,

119-120, 127, 129, 158, 296, 322, 395
Nitric acid, 78, 80, 87, 90, 96, 109, 123-124, 127, 130, 146, 150, 156-160, 171, 227-231, 235, 399
  metals and alloys vs, 228-230
  nonmetallics, 230
  storage of, 68, 140, 161, 218, 234, 236, 258-261, 287-288, 398
Noble metals, 56, 123, 139, 155, 161, 230, 235, 241, 249, 254, 260, 265, 271, 281, 304
Nonaqueous environments, 342-343
Nonmetallic, 3, 72, 93-94, 96, 163, 175, 230, 232, 235, 241, 245, 254, 257, 260, 265, 365, 369
Nonmetallic barriers, 365, 369
Nonmetallic materials, 3, 93, 96, 163, 175, 230, 232, 235, 241, 245, 254, 257, 260, 265
Normalizing of steel, 116
Normal hydrogen electrode (NHE), 61-62, 395
Notch sensitivity, 27-28, 165

Ohm's law, 64, 371
On-stream inspection (OSI), 50, 381-382, 384-385, 387, 395
Optimum materials construction corrosive service, 17, 45, 49
Once-through, 191, 197-199, 201, 203, 208, 319, 340, 398
Organic acids, 78, 109, 143, 146, 156-157, 159-160, 166, 175-176, 239, 253, 256, 258-261, 326, 342
Oxidizing salts, 291, 293
Oxygen concentration cell, 59-60, 70

Paint, definition of, 49, 89, 221-222, 225, 269, 339, 345, 347-351, 358-364
Paints and coatings, 3, 9, 18, 24, 40, 43, 88-89, 96, 110, 145, 147, 161, 163, 169, 171-172, 175, 178, 205, 212-213, 215, 221-

222, 225, 318, 320, 345-350, 352-359, 361-362, 364-365, 369, 376, 379, 390, 398
  application of, 2, 34, 40, 43, 50, 68, 140, 269, 301, 317, 335, 363-364, 368, 371, 383
  economic justification for, 363-364
  surface preparation for, 363
  types of, 5, 7, 27, 32, 38-40, 42, 45-47, 49, 69, 80, 84, 86, 88, 96, 111, 113-114, 124, 145, 147, 149-150, 165, 171, 180, 189, 194, 209-210, 215, 217, 220, 225, 246, 304-305, 321-322, 342, 347-348, 351-352
Passivity, 64, 123-124, 156, 254, 259, 279
Pay-out period (POP), 329, 395
Pearlite, 116
Perfluoroalkoxy, 169
Personal information management (PIM), 390, 395
Phenolphthalein alkalinity, 184, 395
Phenomena-related, 96
Phenolic resins, 281
Phosphoric acid, 78, 136-137, 176-177, 239, 246-248, 262, 283, 342, 398
  metals vs, 247-249
  nonmetals vs, 249
  storage of, 68, 140, 161, 218, 234, 236, 258-261, 287-288, 398
Physical properties in materials selection, 23, 30, 113
Piping, 7, 18, 30-31, 36, 47, 49, 80, 98, 101, 107, 147, 166-168, 172-173, 176, 179, 198, 203, 213, 225, 231, 234, 236, 240, 242, 246-247, 249-250, 256, 259, 261, 268, 276, 278, 280, 282, 293, 307, 311, 321, 371, 377-379, 382, 384, 387
Pitting, 40, 42, 46, 68, 70, 73, 78, 94-96, 98, 101, 109-110, 127, 129, 131, 134-136, 141, 202-203, 205-206, 210, 213, 217, 221-222, 240, 244, 249, 251, 267-269, 271, 273, 291-294,

320-321, 326, 340, 342, 356, 385
Plastics, 8-9, 49, 83-84, 90-91, 96, 147, 160, 163-165, 167-168, 170, 173-175, 180, 207, 214, 232, 235, 239, 242, 245, 249, 255, 257, 260, 265, 275, 282, 286, 293, 350, 369
  thermoplasts, 163-164
  thermosets, 163, 169
Plating, 62, 100, 139-140, 146, 160-161, 223, 232, 365-366, 394
Platinum, 57, 63, 123, 156, 161, 230, 235, 241, 245, 249, 254, 257, 271, 281, 286
Polarization admittance instanteous rate (PAIR), 100, 395
Polarization, anodic, cathodic, 58
Polybenzimidimiazole (PBI), 170
Polycarbonates, 90, 170
Polychlorinated bithenyl (PCB), 17
Polyethylene, 166, 174, 245, 255, 260
Polypropylene, 90, 166, 207, 255, 257
Polytetrafluoroethylene (PTFE), 101, 168-169, 176, 224, 230-231, 235-236, 242, 246, 249, 256, 261, 275-276
Polythionic cracking, 270
Polyvinyl chloride (PVC), 143, 168, 207, 231, 235, 242, 249, 275, 282, 293, 393, 395
Polyvinylidene chloride (PVDC), 168, 233, 235, 255
Polyvinyl fluoride (PVF), 169
Polyvinylidene fluoride, 169
Portland cement, 178, 180, 292
Potable water, 49, 78, 151, 191, 193-194, 206, 340, 352
Potassium hydroxide, 277, 283
Poultice corrosion, 205
Precious metals, 155, 160, 232, 245, 257
  (See also Noble metals)
Precipitation-hardened, 107, 128, 140
Predictable saturation index (PSI), 25, 99, 108, 155-156, 159, 186, 194-195, 264, 395
Present worth (PW), 330-333, 335, 395

Present worth after taxes (PWAT), 333-335, 395

Procedures communications materials selection, 47, 51

Pumps, 9, 17, 32, 42, 47, 49, 101, 111, 140, 168, 171, 173, 176, 189, 198, 203, 231, 234, 236, 242, 244, 246-247, 250, 256, 258, 261, 276, 280, 283, 288, 387

Pyrophoric corrosion products, 159, 241, 254, 272

Quality assurance tests, 80, 86, 93, 96, 368

Quality control heat exchanger tubes, 325-326

Radiography, 383

Ranking alloys, 136

Raw water, 191, 197

Reactive metals, 95, 155, 230, 232, 235, 241, 244, 248, 251, 254, 257, 260, 271, 275, 281, 286, 306, 368

Recycle streams, 48

Reducing acids, 94, 123, 129, 141, 156-157, 159, 227-228, 233, 239

Reduction, 26, 56, 58, 63, 86, 89, 100, 120, 139, 150, 211, 227, 239, 250, 293, 295-296, 304, 325, 329, 371, 376, 397

electrochemical, 3, 10, 43, 62, 64, 68, 72, 84, 88, 90, 97, 100, 105, 123-124, 139, 147, 165, 215, 221-222, 295, 317, 320, 371-372, 384

high-temperature, 3, 20, 24, 35, 69, 120, 128, 133, 155, 170, 197, 295-296, 301, 303, 307

Reduction in area, 26

Refractories, 177

Refractory metals, 155, 232

Refrigeration brines, 341

Regulations, 29, 31, 347, 352

Reinforcement of resins, 170, 282

Reinforced thermosetting plastic (RTP), 172, 395

Reliability and safety, 1, 15, 17, 19, 29, 46, 50, 193, 206, 233, 243, 265, 348, 357, 363, 368, 378, 381

Relative humidity (RH), 218-219, 319, 395

Residual, 33, 59, 84, 86, 94, 114, 117, 149, 196, 200, 206, 240, 279, 312, 322, 340, 366

Resistivity, 210-212, 371, 374, 377

Resources, 4, 20-21, 46, 363

Return on investment (ROI), 329, 395

Rotating equipment, 42, 47, 85, 365, 387

Rubber, 95, 168, 173-174, 177, 207, 231, 242, 245, 255-256, 260, 272, 275, 349, 369

fluorinated, 161, 169, 175, 232, 235-236, 245, 255, 257, 260, 275, 293, 369

natural, 3, 7, 56, 58, 64-65, 107, 109, 123-124, 140, 147, 150, 155, 173, 178, 189-190, 195, 205, 207, 219, 245, 263-264, 267, 275, 312, 346-348, 363, 371, 385

synthetic, 123, 173-174, 191, 219, 245, 260, 346-348

Rust preventatives, temporary, 222, 346

Ryznar Stability Index (RSI), 184-186, 189, 395

Safe ending heat exchanger tubes, 327

Salts, 3, 16, 66, 86, 88, 123, 141, 147, 152, 161, 175, 178-179, 183, 186, 189-190, 192, 194-196, 199-200, 203, 205, 209, 218-219, 223, 227, 239, 251, 253, 258, 270, 277, 283, 285, 291-293, 301, 305, 312, 326, 345, 349, 354, 386

acids, 3, 78, 80, 94, 109-110, 123-124, 129-130, 140-141, 143, 145-146, 150, 153, 156-157, 159-161, 166-168, 171-173, 175-176, 180, 187, 227-229, 231, 233, 235, 239, 245, 253-

254, 256, 258-261, 267, 270-271, 273, 277, 283, 292-293, 319, 326, 342, 348-349, 351-352, 355
alkaline salts , 16, 90-91, 109, 161, 177, 179, 195, 222-223, 225, 277-278, 283, 285, 292-293, 305, 318-319, 351, 354
neutral, 105, 135, 177-178, 291-293, 356
oxidizing, 16, 56, 78, 94, 123-124, 129-130, 134, 140-143, 146, 150, 153, 156, 158-161, 167, 172, 174-175, 179-180, 191, 199, 204, 227-228, 230-231, 233, 235, 239-241, 248-255, 257, 259-260, 273, 279, 281, 286, 291, 293, 304-305, 307, 320, 340, 343, 347, 355, 386
Saturated Calomel Electrode (SCE), 63, 376, 395
Scaling, 183-186, 192, 199
Scaling indexes in corrosion, 184-186
aggressiveness index (AI), 186
by water, 183, 185, 187, 205, 209
Scanning electron microscope (SEM), 70, 152, 385-386, 395
Scavenging corrosive species, 318, 320
Seal welds, 323
Seawater, 63-67, 78, 109, 130, 140, 143, 151, 153, 156, 172, 179, 187, 189-193, 197-198, 205-208, 210, 260, 319, 340, 373, 376, 378
Sensitization, 77-80, 94, 259, 270, 302
Service-related, 97, 386
Shielded metal arc welding (SMAW), 35, 396
Shot-peening, 84-85, 322
Silicon, 25, 32, 34, 67, 74, 86, 89, 108, 111-114, 119, 130-131, 139-140, 151-152, 174, 177, 195-196, 223, 229, 232, 234-235, 243, 247, 256, 258, 305-306, 350, 378
Silver, 16, 24, 56-57, 62-63, 123, 145, 160-161, 230, 232, 235, 241, 245, 249, 254, 257, 260, 271-

272, 281, 286, 298, 305, 376
Single-metal corrosion, 57, 59
Slow strain rate embrittlement (SSRE), 268-269, 295-296, 396
Slushing compounds, 346
(See also Rust preventatives, temporary)
Sodium hydroxide, 140, 195, 277-280, 283, 291, 398
environmental cracking by, 312
metals vs, 140-141, 278-281
nonmetals vs, 281-282
storage of, 68, 140, 161, 218, 234, 236, 258-261, 287-288, 398
Soil, corrosion by, 210
chemistry of soil, 210-211
composition and condition of soil, 211
soil, 3, 49, 58, 65, 68, 209-215, 339, 371-374, 376-378
corrosion control in, 1, 365, 398
factors in, 3, 23, 46, 210
resistivity, 210-212, 371, 374, 377
specific materials in, 211-214
types of corrosion in, 209
Soldering, 34-35, 43
Special effects in corrosion, 18
(See also Hazardous materials)
Specific equipment in materials selection, 46
Specific gravity, 23
Specification in materials selection, 208
Stabilization of stainless steels, 79
Stainless steels, 8, 10, 18, 24, 27, 31, 34, 39, 42, 50-51, 64, 72-74, 77-81, 83, 88-89, 94-96, 102, 121, 123-130, 132-134, 137, 139, 142, 198, 206, 213, 218, 223, 225, 229-230, 232, 234, 240-241, 244, 247, 251, 257-258, 264-265, 268, 270, 273, 278-280, 286, 288, 292, 294, 302, 307-308, 312, 319, 342, 368, 384, 398-399
alloy designations for, 107
austenitic, 10, 24, 34, 42, 50-51, 68, 72-73, 77, 81, 83, 86-88, 94-

96, 102, 113, 125, 127-130, 133, 137, 142, 164, 198, 213, 224, 229, 234, 247, 251, 258-259, 268, 270, 272, 279, 298, 302, 307-308, 312, 319, 323, 341, 367, 384
castings, 4, 32, 34-35, 46, 107, 109, 112, 131, 149-150, 229, 308
corrosion of, 2, 18, 55, 60, 65-67, 71, 109, 132, 143, 159, 162, 178, 186-188, 205, 209, 213, 218, 223-227, 229, 239, 250-253, 259, 261, 264, 269, 292, 294, 309, 313, 319, 323, 342-343, 345-346, 371, 383
duplex, 34, 112, 121, 128-131, 198
ferritic, 27, 74, 80, 87-88, 124-127, 229, 251, 258, 270, 278
high-performance, 128-129
martensitic, 125-126
nature of, 15, 21, 25, 27, 77, 84, 105, 109, 123, 190, 210, 213-214, 218, 227, 257, 287, 304, 317, 339
passivity, 259
precipitation-hardened, 107, 128, 140
specialty, 9, 118, 174, 234, 308, 378
superferritic, 18, 27, 127, 129, 206, 323
types of, 5, 7, 27, 32, 38-40, 42, 45-47, 49, 69, 80, 84, 86, 88, 96, 111, 113-114, 124, 145, 147, 149-150, 165, 171, 180, 189, 194, 209-210, 215, 217, 220, 225, 246, 304-305, 321-322, 342, 347-348, 351-352
Standard half-cells, 61
Standard hydrogen electrode (SHE), 61, 63, 376, 395
Standards in materials selection, 5, 7
Steam, corrosion by, 20, 29, 40, 48-49, 65, 72, 87-90, 142, 152-153, 174, 183, 192, 194, 196-197, 204-205, 207, 263-264, 280, 296, 304, 307, 325, 341, 343, 345-346, 359
(See also Water, corrosion by)

Steam condensate, 49, 65, 90, 192, 197, 204-205, 264, 343, 345
Steels, 8, 10, 18, 24-25, 27, 31, 33-34, 39, 42, 50-51, 64, 72-74, 77-81, 83, 85, 87-89, 94-96, 102, 111-112, 114-115, 117-121, 123-130, 132-134, 137, 139, 142-143, 197-198, 204, 206, 212-213, 218, 221-223, 225, 229-230, 232, 234, 240-241, 243-244, 247, 251, 257-258, 264-265, 268, 270, 273, 278-280, 286, 288, 292, 294, 296-297, 299, 302, 306-308, 312, 319, 342, 368, 384, 398-399
carbon, 27, 36-37, 67, 72, 77-80, 87, 107, 111-112, 114-116, 118-121, 124, 127-129, 131, 134-135, 140, 142, 159, 164, 170-171, 173-176, 183, 192-193, 197, 203-204, 218-219, 222, 229-230, 235-236, 239, 246, 249, 255-257, 259-260, 263-265, 268, 273, 275, 278, 281-282, 285-288, 297, 304, 306, 323, 341, 343, 350, 366-368, 378
heat treatment of, 94, 112, 121
numbering system for, 31, 121
weathering, 204, 221-222, 345, 348, 350-351, 354
Step-wise cracking, 268, 396
Stiff-Davis index (SDI), 186, 395
Stress corrosion cracking (SCC), 73, 83-90
(See also Environmental cracking)
Stress relief: 18, 24-26, 28, 30, 33-34, 41, 56, 59, 68, 73-74, 79, 83-85, 141, 312, 322, 368
mechanical, 2, 4, 6, 15, 18, 23, 25-27, 29, 31, 33-34, 40-43, 47, 55, 69, 72, 84-86, 90, 98, 107, 111-113, 115-116, 118, 120, 128, 139, 143, 145-146, 150, 156, 158, 163-166, 169-171, 176-177, 213-214, 234, 255, 275, 301-302, 304, 307, 312, 321-323, 359, 368, 382, 390

thermal, 19, 24, 33, 41, 78-79, 84-85, 94-96, 112-113, 129, 141, 164-166, 170-171, 175-177, 197, 209, 215, 219, 224, 226, 234, 278, 322-324, 358, 367-368

Stress-strain diagram, 26

Stresses, 24, 28, 33, 41-42, 83-86, 98, 117, 178, 224, 278, 302, 323, 327, 342, 366-368

applied, 18, 24, 26, 28, 33-34, 67, 84, 94, 98, 169, 173, 193, 225, 278, 312, 320, 322, 345, 347-350, 353, 355, 357, 361-362, 365, 377, 383, 387, 389

compressive, 84, 322

in corrosion testing, 94

residual, 33, 59, 84, 86, 94, 114, 117, 149, 196, 200, 206, 240, 279, 312, 322, 340, 366

tensile, 25-26, 42, 73-74, 83-84, 86, 107-108, 115, 120, 155, 159, 164, 296, 312, 322

Stress-oriented hydrogen-induced cracking (SOHIC), 268, 396

Submerged arc welding (SAW), 36, 395

Subsurface waters, 189

Sulfate bacillus, 179

Sulfate-reducing bacteria (SRB), 199, 203, 214, 271, 376, 396

Sulfidation, 305

Sulfide stress cracking (SSC), 28, 56, 73, 83, 85, 91, 203, 210, 268-271, 288, 396

(See also Environmental cracking)

Sulfuric acid, concentrated, 68, 130, 160, 175, 228-229, 233-235, 237, 239, 250-251, 398

vs metals, 251-254

vs nonmetals, 254-255

storage and handling of, 255-256

Superaustenitics, 134

Superferritic stainless steels, 27, 127

Surface finish materials selection, 42

Surface stability, 303

Tantalum, 32, 155, 158-160, 177, 228, 230, 235, 241, 244, 249, 254, 257, 260, 271, 275, 281, 286, 293, 298, 366

Technical service department as a resource, 4

Technical societies as a resource, 5

Telluric currents, 209, 212

Tempered water, 193

Tempering, 33, 116, 302

Tensile strength, 25-26, 86, 107, 115, 155, 159

Terneplate, 147, 223

Testing, corrosion, 3, 5, 46, 50, 93, 95, 101-102, 129, 326, 344, 399

Thermal conductivity, 24, 175-177

Thermal expansion, 24, 84, 113, 164, 170, 215, 323, 367-368

Thermal history, 79, 94, 96

effect of, in corrosion testing, 59, 68, 73, 84, 86, 108, 151, 197, 223, 230, 291, 371, 374-375

Tin, 57, 63, 105, 145, 147, 149, 151-152, 158, 232, 257, 274, 311

Titanium, 37, 63-64, 66, 78-79, 87, 90, 134, 155-159, 162, 207, 218, 224, 228, 230-231, 235, 241, 244, 248, 254, 257, 260, 265, 271, 273, 275-276, 281, 293, 298, 323, 378, 384

corrosion of, 2, 18, 55, 60, 65-67, 71, 109, 132, 143, 159, 162, 178, 186-188, 205, 209, 213, 218, 223-227, 229, 239, 250-253, 259, 261, 264, 269, 292, 294, 309, 313, 319, 323, 342-343, 345-346, 371, 383

grades and UNS numbers of, 157

Total dissolved solids (TDS), 184, 191, 396

Total hardness (TH), 184, 192, 396

Toughness, 26-27, 33, 50, 116, 119-120, 157, 167-168

Trade associations, 7

Tube-to-tubesheet joints, 323

Two-metal corrosion, 47, 70

Types of waters, 189

Ultrasonic inspection, 74, 368
Ultraviolet (UV), 90, 166-167, 173, 191, 348, 350, 352-355, 358, 396
Unalloyed, 111, 156-158
Underground, 18, 49, 67-68, 107, 172, 198, 212-215, 349, 353, 371, 374, 377, 379
Unified numbering system (UNS), 2, 10, 17, 25, 28, 31-32, 37, 43, 74, 77, 79, 97, 107-108, 121, 126, 130-131, 133-137, 139-141, 143, 147, 149-150, 152, 157, 206, 224, 229-230, 232, 234-235, 241, 243-244, 247, 251, 256-258, 265, 270, 274, 279-281, 286, 288, 308, 312, 318, 366
Uniform (general) corrosion, 69-70, 73, 87, 94-95, 127, 131, 145, 217, 244, 267, 293-294, 311, 321-322, 340, 346, 368, 387
Unscheduled shutdowns, 16, 46, 381

Valves, 9, 17, 32, 47, 49, 101, 111, 140, 146-147, 160, 168, 170-171, 173, 176, 198, 203, 231, 234, 236, 242, 244, 246-247, 250, 253, 256, 261, 276, 280, 283, 288, 321
Velocity effects, 47, 69, 72, 97, 234, 267
Videotape programs, 399
Vinyl chloride-monomer (VCM), 17
Visual examination, 70, 382, 385, 391
Volatile organic compounds (VOC), 347-348, 352, 356, 359, 396

Water, corrosion by, 2-3, 8, 16, 19-20, 23, 48-50, 56-58, 60, 66, 68, 72, 78, 80, 86-89, 105, 109, 117, 123-124, 126-127, 141, 147-148, 151, 156-158, 166-168, 172, 174, 178-179, 183-187, 189-194, 196-209, 211, 214-215, 217-218, 220-221, 224-225, 227, 229-230, 239, 242, 245, 251-252, 256, 263, 265, 267-268, 270, 273, 275, 277, 280, 283, 285-287, 291-293, 305, 312, 317, 319, 322, 325-326, 339-341, 343, 346, 348-349, 351-355, 357, 359-361, 371-372, 378, 393, 395, 398
boiler feedwater (BFW), 188, 191-192, 393
brackish water, 189-190
chemistry of water, 183
chilled water, 193
condensate, steam, 194-197
cooling water, 19-20, 48-49, 60, 191, 193, 197, 199, 319, 326, 340
corrosion of steel, 18, 66, 187, 218, 224-225, 251-252, 319, 342, 383
cycle water, 199-201
demineralized water (DM), 191-192, 252, 393
desalinated water, 193
dissolved oxygen (DO), 56, 58, 66, 150, 184, 186-187, 248, 252, 261, 264, 267, 292, 319, 339, 393
distilled water, 72, 190-192, 252, 263, 312
fire-control water, 49, 193
fresh water, 189-191, 197-198, 205, 207, 339
hardness of waters, 184-186
inhibition, 201
metals, 204-207
nature of waters, 189
origin of waters, 189
pitting in, 78, 206, 271, 356
plastics and elastomers, 207
potable water, 49, 78, 151, 191, 193-194, 206, 340, 352
process water, 193
produced, water, 189, 193
raw water, 191, 197
scaling indices, 184-186
seawater, 63-67, 78, 109, 130, 140,

143, 151, 153, 156, 172, 179, 187, 189-193, 197-198, 205-208, 210, 260, 319, 340, 373, 376, 378

special information sources in, 207

specific materials,187-188, 204-207

stress corrosion cracking (SCC) in, 18, 24, 68, 73, 83-84, 88-89, 91, 102, 129, 162, 226, 272, 326-327, 393-395

subsurface waters, 189

surface waters, 189-190, 197

tempered water, 193

types of, 189

uses of, 190

Weld decay, 72, 77-80, 95, 241, 270

Weld defects, 40, 367

Weld overlays, 318, 366-367

Welding, 7, 10, 17, 25, 30, 33-38, 41-43, 50, 67-68, 74, 77-79, 84, 86, 88-89, 95, 99, 107-108, 117, 126-129, 141-142, 149, 152, 155-156, 158-159, 166, 168-169, 213, 228-229, 259, 298, 305, 318, 321, 323, 325, 366-367, 394-396, 398-399

White iron, 111-112

Wood, 179-180, 246, 260, 348

Wrought products, 33

X-ray diffraction, 386

X-ray inspection, 382-383

Yield point, 25-26, 225, 278

Yield strength, 26, 28, 107-108, 298, 301

Zinc, 24, 57, 63-64, 66-67, 70, 73, 86, 89, 105, 108-110, 145, 147, 151, 173, 205-206, 213, 220-223, 227, 257, 264, 269, 271, 274, 277, 298, 339, 341, 351, 356, 366, 372-373

Zirconium, 15, 80, 90, 150, 155, 158-

159, 207, 228, 230, 235, 241, 244, 249, 254, 256-257, 260, 271, 275, 281, 298

Notes